DATE DUE FOR RETURN

Successful Women in Chemistry

Anne Leslie is a member of the ACS WCC and currently operates a craft business specializing in paper cutting arts. In honor of this book, she has provided this illustration showing both the scientific and corporate sides of successful women in chemistry today.

Anne is a retired biochemist, having worked for the U.S. EPA Pesticides Office for 18 years. Her expertise was in Integrated Pest Management. She published a book in 1994 with CRC Press: *Integrated Pest Management for Turgrass and Ornamentals*. It included chapters by experts from academia, industry, and government and was organized and edited by Anne. It was considered a definitive collection on the subject and was used in coursework in the United States and a number of other countries, including New Zealand and Australia. Anne is active as a Councilor in the Chemical Society of Washington and serves on the National Project SEED Committee as well as the WCC. She holds an M.S. in Biochemistry from McGill University and a Ph.D. (all but dissertation) from the University of Utah.

ACS SYMPOSIUM SERIES **907**

Successful Women in Chemistry

Corporate America's Contribution to Science

Amber S. Hinkle, Editor
Bayer Corporation

Jody A. Kocsis, Editor
Lubrizol Corporation

Sponsored by the
ACS Women Chemists Committee,

American Chemical Society, Washington, DC

Library of Congress Cataloging-in-Publication Data

Successful women in chemistry : corporate America's contribution to science / Amber S. Hinkle, editor, Jody A. Kocsis, editor ; sponsored by the ACS Women Chemists Committee.

    p. cm.—(ACS symposium series ; 907)

    Includes bibliographical references and index.

    ISBN-13: 9780841239128

    ISBN 0-8412-3912-6 (alk. paper)

    1. Women chemists—United States—Biography. 2. Chemists—United States—Biography. 3. American Chemical Society. Women Chemists Committee.

    I. Hinkle, Amber S., 1968- II. Kocsis, Jody A., 1966- III. American Chemical Society. Women Chemists Committee. IV. Series.

QD21.S78   2005
540′.92′273—dc22   2005043612

The paper used in this publication meets the minimum requirements of American National Standard for Information Sciences—Permanence of Paper for Printed Library Materials, ANSI Z39.48–1984.

Copyright © 2005 American Chemical Society

Distributed by Oxford University Press

PRINTED IN THE UNITED STATES OF AMERICA

Foreword

The ACS Symposium Series was first published in 1974 to provide a mechanism for publishing symposia quickly in book form. The purpose of the series is to publish timely, comprehensive books developed from ACS sponsored symposia based on current scientific research. Occasionally, books are developed from symposia sponsored by other organizations when the topic is of keen interest to the chemistry audience.

Before agreeing to publish a book, the proposed table of contents is reviewed for appropriate and comprehensive coverage and for interest to the audience. Some papers may be excluded to better focus the book; others may be added to provide comprehensiveness. When appropriate, overview or introductory chapters are added. Drafts of chapters are peer-reviewed prior to final acceptance or rejection, and manuscripts are prepared in camera-ready format.

As a rule, only original research papers and original review papers are included in the volumes. Verbatim reproductions of previously published papers are not accepted.

ACS Books Department

Contents

Preface...xi
    Madeleine Jacobs

Editors' Comments..xiii

1.   **Women in the Chemical Professions: An Overview**...............................1
    Frankie Wood-Black and L. Shannon Davis

2.   **Susan B. Butts, Director of External Technology: Major
    Chemicals Manufacturing**..11
    Arlene A. Garrison

3.   **Anne DeMasi, Regulatory Specialist, A Family-Oriented
    Career Path**..17
    Jacqueline Erickson

4.   **L. Shannon Davis, Leader of Process Research and
    Development: Adventures of an Industrial Chemist**............................21
    L. Shannon Davis and Amber S. Hinkle

5.   **Sally Sullivan, Patent Attorney: Combining Chemistry and Law**.......31
    Ellen A. Keiter

6.   **Louise B. Dunlap, Director of Technology Transfer: A Career
    in Creating Partnerships**...37
    Arlene A. Garrison

7.   **Nancy Jackson, Deputy Director: A Full-Circle Career Path**............43
    Ellen A. Keiter

8.   **Sharon V. Vercellotti, Company President: A Business
    of One's Own**..51
    Rita S. Majerle

9. Jean Zappia, Vice President of Plastics Additives Segment:
    Achieving Life Balance..57
    Jody A. Kocsis

10. Rachael L. Barbour, Senior Chemist: Spectroscopy
    for Problem-Solving..63
    Arlene A. Garrison

11. Elizabeth Yamashita, Group Director of Global Regulatory
    Sciences: The Challenge of Evolving Regulations.......................69
    Jacqueline Erickson

12. Lora Rand, Vice-President of Plastics Manufacturing:
    Inspiring Leadership...75
    Amber S. Hinkle

13. Barbara J. Slatt, Manager of Corporate R&D and Corporate
    External Relations: A Broad Spectrum of Opportunities...................79
    Jacqueline Erickson and Elizabeth A. Piocos

14. Cheryl A. Martin, Director of Financial Planning: From the
    Laboratory to Wallstreet...87
    Amber S. Hinkle

15. Rebecca Seibert, Technology Manager: Making Choices....................93
    Jody A. Kocsis

16. Grace Torrijos, Owner and President: Business
    as a Family Affair...101
    Rita Majerle and Elizabeth A. Piocos

17. Diane M. Artman, Marketing Director: Transitions.......................105
    Jody A. Kocsis

18. Marion Thurnauer, Senior Scientist: A View from a
    Government Laboratory..111
    Ellen A. Keiter and Elizabeth A. Piocos

19. Susan E. James, Vice-President of Worldwide Regulatory
    Affairs and Regulatory Compliance: Chemistry around
    the World..117
    Jacqueline Erickson

20. Rita A. Bleser, Vice-President for Research and Development:
A Unique Path to Management...123
Ellen A. Keiter

21. Beverly Dollar, Chief Intellectual Property Counsel: The Joy
in Chemistry...129
Frankie Wood-Black

22. Sharon L. Haynie, Research Scientist: Laboratory
Investigation and Outreach...135
Arlene A. Garrison

23. Sharon L. Fox, Strategic Planning Consultant: Working
in a Truly Global Capacity..139
Amber S. Hinkle

24. Tanya Travis, General Manager: Building Confidence....................147
Jody A. Kocsis

25. Melissa V. Rewolinski, Director of Chemical Research and
Development: A Career in Teamwork and Leadership......................153
Rita S. Majerle

26. Margaret A. (Lissa) Dulany, Chemical Consultant and Writer:
Chemistry and People..159
Amber S. Hinkle and Margaret A. (Lissa) Dulany

27. Shirlyn Cummings, a Human Resources Perspective: Leaders
for a Global Market...165
Amber S. Hinkle

28. Lessons Learned..173
Arlene A. Garrison

Meet The Authors..179

Indexes

Author Index...189

Subject Index..191

Preface

When I speak to young women about how to be successful, I give them several pieces of advice: Believe in yourself, find a mentor, talk to women who are successful and find out how they did it, and, above all, find out what your passion is and follow it.

With the publication of this book, I will have to add another tip: Read this book! Inside these pages, you will find the rewarding stories of women chemists who have made it to the top in corporate America and in government laboratories. Their success stories resonate with inspiration, wise advice, and humor. I knew many of these women before reading the book, but I learned something new even about those women from reading their stories. I hope readers of this book will as well. More importantly, I hope readers will give this book to girls who they might want to inspire to become chemists. The stories of these women should indeed inspire our next generation.

Above all, I hope that this book will convince all young women who aspire to be at the top of their profession that, as the famous lyricist and songwriter Cole Porter once wrote, "The apple on the top of the tree is never too high to achieve." But to get that apple, women must possess self-confidence.

A lack of self-confidence and a lack of self-esteem are nearly universal problems that haunt women at some stage in their development. Girls and young women in particular suffer from a lack of self-confidence that holds them back. Still, this is not limited to youth. All of us know women well into their 40s who are very successful at what they do who still think they aren't good enough, smart enough, thin enough, or pretty enough to succeed.

It's possible that men have a gene for self-confidence on their Y chromosome, because research has shown that the Y chromosome has been shedding genetic material for years and there is not a lot left. I don't really believe this is true, but sometimes women do act as though they are missing that gene for self-confidence. I don't think we can wait for cloning to solve this problem. We must help each other to succeed by reinforcing how very good we are!

All of the women profiled in this book not only have self-confidence (or learned to gain it), but they also have a passion for what they do. And this is my last bit of advice for all who aspire to be successful, and the women in this book exemplify this advice: No matter what career pathway you choose, you should do it for something other than money. You should do it because you love it with a passion, and because you believe in it.

A career is like a love affair. It has its ups and downs--but overall, it must be rich and rewarding. It must be bursting daily with possibilities and promise, or why bother? It must provide an environment in which you can constantly grow and learn. It must make you want to get up every day and go to work. It must be fun.

I can't imagine staying in a relationship that didn't have these qualities, and what else is a job except a relationship where you spend anywhere from eight to 16 hours a day? I can't imagine spending even a minute at work without feeling passionate about it.

It's not always easy to find the right person—or the right career. Sometimes we simply don't choose wisely the first time; presumably, we learn from that experience and move on. Making that transition can be painful, but it can also open up exciting possibilities. So it is with a career. The important thing to understand is that there are alternatives— that there might not be a single right choice that will give a lifetime of satisfaction. Fortunately, there's a veritable cornucopia of career possibilities for people trained in chemistry at all degree levels.

I am a firm believer in seizing the day. That is what the role models in this book did, and I urge the readers of this book to use every day to advance your careers. As the famous public speaker Dale Carnegie advised: "Today is the first day of the rest of your life." Make the most of it.

Madeleine Jacobs
Executive Director and Chief Executive Officer
American Chemical Society
1155 16th Street NW
Washington, DC 20036

Editors' Comments

On a roller-coaster ride, one may experience ups, downs, thrills, excitement, and—in some cases—fright; the same can be said for both personal and professional experiences. This book is intended to reach a variety of people, allowing them a glimpse of what others have done with their science backgrounds and to gain an understanding of the multitude of options available.

Based on a series of interviews, the book showcases more than twenty women chemists and their compelling success stories. In highly readable and honest accounts, these women tell of the challenges, positive influences, and personal aspirations that have helped shape their individual careers. The diverse professional choices they have made and the range of innovative strategies they have employed guarantee their stories will provide inspirational reading for anyone interested in achieving success in chemistry—or any technical field.

The first chapter includes information on the general status of women in the chemical industry, while Chapters 2 through 26 capture the women's success stories. At the end of this book, readers are given the perspective from corporate human resources, a summary of the pervasive messages found in interviewing these women, and the opportunity to read about the authors themselves.

This book features several women from the American Chemical Society (ACS) Women Chemists Committee (WCC) newsletter series, *Successful Women in Chemistry,* as well as from the 2003 symposium *Corporate America's Contribution to Science: Successful Women Chemists,* held at the 225th National Meeting of the American Chemical Society.

We thank the authors, interviewees, symposium sponsors, and Cheryl Brown for administrative assistance. This has been a wonderful experience. We hope you enjoy and learn from this book as much as we have enjoyed interviewing these women who have made significant inroads into science.

Amber S. Hinkle

Bayer Materials Science
8500 West Bay Road
Mail Stop 18
Baytown, TX 77520

Jody A. Kocsis

Lubrizol Corporation
29400 Lakeland Boulevard
Wickliffe, OH 44092

Chapter 1

Women in the Chemical Professions: An Overview

Frankie Wood-Black[1] and L. Shannon Davis[2]

[1]ConocoPhillips, 6855 Lake Road, Ponca City, OK 74604
[2]Solutia, Inc., 3000 Chemstrand Road, Cantonment, FL 32560

The numbers of women choosing careers in the chemical professions have risen dramatically over the past 20 years. Yet, the corresponding rise in middle and senior levels of all career paths has not materialized. An overview of the current demographics of women in the chemical professions and potential causes for this discrepancy is presented.

Women provide leadership in the chemical sciences and make significant contributions to the understanding of the chemical sciences. This symposium was devoted to highlighting women in mid to upper level positions within the chemical industry who have been deemed "successful." Success comes in many forms and via many career paths. This book focused on women from industrial and government backgrounds because the authors wanted to show how science impacts the corporate world and what roles women have and will continue to play. The book highlights women from a variety of careers, ranging from very technical, obvious careers to those that are not as technical or as obvious.

The past 20 years have shown a dramatic change in workforce demographics in the chemical professions, as illustrated in the American Chemical Society (ACS) workforce survey data. Since the 1970s, women and minorities have slowly increased their presence in the chemical professions. The pipeline theory has ostensibly worked as planned, as evidenced by the numbers of women and minorities obtaining degrees in chemistry. However, while the pool of candidates has increased, the numbers of women and minorities in the chemical workforce have not increased in similar proportions. Based on a 2002 survey by *Catalyst*, less than 16% of corporate officers are female, almost double the 9% reported in 1995[1]. Women represent only 12% of the faculty at the top 10 National Science Foundation (NSF) research schools; only 7% of the full professors are female[2]. These observations lead to a variety of questions such as:

- What are the root causes that lead to a slower assimilation of women in the higher echelons of the chemical professions?
- Are the issues systemic?
- Is the pipeline theory really working, and if not, why not?
- Do any significant factors make the chemical professions unfriendly to women?

These issues have intrigued scholars, study groups, and the Women Chemists Committee (WCC) of the ACS for many years.

Many people may be asking why this book and why this book now. These are interesting questions. The answer depends on exactly where you are in the career cycle. It depends upon your experience and your perspective. For some, the issue is a legacy of the past—the numbers of women in the chemical professions are increasing; women appear to have solved all of the problems facing them; and the pipeline is full, hence it is just a matter of time before equity is achieved. For others, the issue is an ever-present barrier and roadblock where the data and numbers do not match, the opportunities do not exist, and equity is a far-off dream. So, what exactly is reality?

As you might expect, reality falls in-between and in the eyes of the beholder. A significant debate continues to rage about exactly what forces come into play—are they societal, is it workplace culture, is it an over-riding issue within the science and engineering workforce, or is it a cultural difference between men and women? What we do know is that, although progress has been made over the past 25 years of data collection, shortcomings remain: Women are not as well recognized as their male counterparts; women on corporate boards of directors are still rare; and women chief executive officers, chief operating officers, and chief technical officers are still newsworthy items. So what is the current status?

The Science and Engineering Workforce

The National Science Board (NSB) recently published "The Science and Engineering Workforce—Realizing America's Potential"[3]. In this report the NSB highlights two themes that have been prevalent for the past 10 years:

- Global competition for science and engineering talent is intensifying, such that the United States might not be able to rely on the international science and engineering labor to fill unmet skill needs.
- The number of native-born science and engineering graduates entering the workforce is likely to decline unless the nation intervenes to improve success in educating science and engineering students from all demographic groups, especially those that have been underrepresented in science and engineering careers.
- In short, the United States is facing a decline in the number of students entering science and engineering fields. Yet science and technology are the engines of economic growth, and a scientifically literate public is necessary to continue providing the discoveries and innovations that are needed to sustain the standard of living that we have all come to enjoy. Eventually, society will be faced with some difficult challenges—energy, sustainability, environmental impacts, to name just a few—which only those well-versed in the fundamentals of science will be equipped to pursue and solve. Thus, the NSB and the NSF have been sounding the alarm: We are not producing enough of the raw material, i.e., science and engineering graduates from the traditional sources. If we believe that this is of national importance, we must look to underrepresented populations to make up the difference, i.e., minorities and women.
- To illustrate the point let's look at some of the detailed statistical information provided by the NSF[4]:
- Science and engineering jobs increased by 159% between 1980 and 2000 to more than 3.6 million nonacademic science and engineering occupations. Yet, 67% of those with science and engineering baccalaureates or higher-level degrees in the workforce in 1999 were in occupations not formally classified as science and engineering jobs but stated that their jobs were at least somewhat related to the their highest science and engineering degree field.
- While the total number of nonacademic science and engineering graduates is rising, the number of those graduates in the physical sciences, which includes chemists, is remaining relatively flat.
- Women and minorities are making significant gains, but this has occurred only recently. Women now make up 24.7% of the science and engineering workforce but 48.6% of the college-degreed workforce. Blacks were 6.9% of the science and engineering workforce but 7.4% of the college-degreed workforce. Hispanics were 3.2% of the science and engineering workforce but 4.3% of the college-degreed workforce.
- Many differences in employment characteristics between men and women are due in part to differences in time spent in the workforce. Women in the

science and engineering workforce are younger on average than men; 50% of women and 36% of men employed as scientists and engineers in 1999 received their degrees within the past 10 years.

The overall characteristics of the science and engineering workforce are changing. Despite indications of progress, some issues must still be addressed. For example, if women are increasingly entering science and engineering fields and have done so in increasing numbers since the late 1970s, where are they in senior and middle management? Where are they in the workforce? What environmental factors are influencing their decisions? Finally, if competition exists for science and engineering trained professionals, why do we still see salary gaps and employment concerns?

Status of Women in the Chemical Professions

The number of women entering the chemical workforce has increased significantly over the past 50 years (Figure 1)[5]. Women in 2000 reported having more bachelors and master's degrees combined than men (55 vs. 34%, respectively). Thus, women are gaining and passing men in degree attainment, yet, men had more doctorate degrees (64.4 vs. 44.7%) and men are more likely to be fully employed than women.

Women are more likely to be employed part-time (5.6% of women vs. 2.1% of men) and are more likely to be not seeking employment (4.1% of women vs. 1.0 % of men). Why are women more likely than men to be employed part-time or not seeking employment? These specific questions were asked during the ChemCensus survey of 2000. When asked the reasons for working part-time, 15.4% of women vs. 23.3% of their male counterparts responded that full-time work was not available and that 36.5% of women vs. 3.9% of men indicated a constraint due to family or marital status. In the case of unemployed chemists not actively seeking employment, 55.7% of women vs. 13.2% of men responded that a family situation was responsible. Also, the ChemCensus data indicated that women tended to put more restrictions on their job search than men. Yet, the ChemCensus data did indicate changes: Women appeared to be considerably less likely to be working part-time due to family constraints than they were in 1990 (Figure 2).

Where Are Women Working?

Women and men are equally represented in all three of the major employment categories: industry, government, and non-academia at the Bachelor of Science level, with slightly more women than men choosing academic careers (6.3 vs. 4.5%). This difference is more marked at the Doctorate level, with

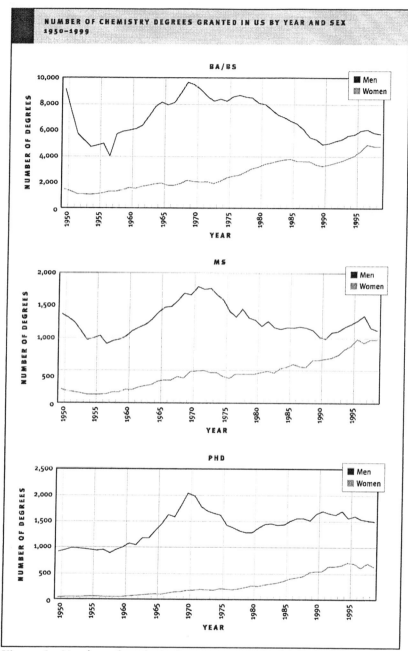

Figure 1. Number of graduate degrees granted in the United States by year and gender, 1950–1999[5].

UNEMPLOYED CHEMISTS ACTIVELY SEEKING AND JOB SEARCH RESTRICTIONS BY SEX									
	ALL CHEMISTS			ALL MEN			ALL WOMEN		
	YEAR			YEAR			YEAR		
ALL CHEMISTS	1990	1995	2000	1990	1995	2000	1990	1995	2000
INABILITY TO RELOCATE	27.1%	21.1%	29.6%	24.6%	16.9%	27.4%	37.3%	34.0%	36.9%
FAMILY RESPONSIBILITIES	6.6	7.2	10.7	5.3	5.9	9.1	12.0	11.3	15.7
NEED FOR PART-TIME EMPLOYMENT	2.9	3.0	3.6	1.7	2.9	3.8	8.0	3.2	3.0
OTHER	10.9	10.8	12.2	11.0	11.3	10.9	10.7	9.3	16.2
NO RESTRICTIONS	52.4	58.0	43.9	57.5	63.1	48.8	32.0	42.1	28.3
TOTAL	100.0%	100.0%	100.0%	100.0%	100.0%	100.0%	100.0%	100.0%	100.0%

RESTRICTIONS ON JOB SEARCH

Figure 2. Unemployed chemists actively seeking positions and job search restrictions by gender[5].

substantially fewer women in industry (43.6 vs. 54.3%) and substantially more in academia (43.9 vs. 34%). These differences are consistent over the 10-year survey period (1990–2000; Figure 3). Given the salary differences between industry and academe, the choice of career path is one factor that contributes to pay inequities between men and women chemists.

			ALL CHEMISTS			ALL MEN			ALL WOMEN		
			YEAR			YEAR			YEAR		
		ALL CHEMISTS	1990	1995	2000	1990	1995	2000	1990	1995	2000
BACHELOR CHEMISTS		INDUSTRY	73.2%	77.9%	83.1%	74.5%	79.8%	83.7%	69.9%	73.8%	81.9%
		GOVERNMENT	10.0	8.0	7.6	9.7	7.7	7.7	10.9	8.7	7.3
		OTHER NONACADEMIC	13.0	5.0	4.2	12.7	4.7	4.1	13.7	5.8	4.4
		ACADEMIC	3.8	9.1	5.1	3.1	7.8	4.5	5.5	11.7	6.3
		TOTAL	100.0%	100.0%	100.0%	100.0%	100.0%	100.0%	100.0%	100.0%	100.0%
HIGHEST DEGREE — MASTERS		INDUSTRY	64.9%	66.3%	69.9%	68.1%	70.6%	73.7%	55.7%	55.9%	62.1%
		GOVERNMENT	10.3	8.9	8.2	10.1	8.6	8.4	11.0	9.5	7.8
		OTHER NONACADEMIC	10.2	5.7	4.9	10.1	5.5	4.2	10.7	6.3	6.1
		ACADEMIC	14.5	19.1	17.0	11.8	15.3	13.6	22.5	28.3	24.0
		TOTAL	100.0%	100.0%	100.0%	100.0%	100.0%	100.0%	100.0%	100.0%	100.0%
DOCTORATE		INDUSTRY	49.1%	49.8%	52.4%	50.3%	51.9%	54.3%	40.2%	38.9%	43.6%
		GOVERNMENT	8.0	7.9	7.1	7.9	7.6	7.0	8.9	9.6	7.6
		OTHER NONACADEMIC	7.1	4.6	4.7	7.1	4.4	4.7	7.4	5.6	4.9
		ACADEMIC	35.8	37.6	35.8	34.7	36.1	34.0	43.5	45.8	43.9
		TOTAL	100.0%	100.0%	100.0%	100.0%	100.0%	100.0%	100.0%	100.0%	100.0%

ALL CHEMISTS BY EMPLOYER, HIGHEST DEGREE AND SEX, 1990–2000

Figure 3. All chemists by employer, highest degree, and gender, 1990–2000[5].

What Is the Family Status?

Family status varies widely by gender. While only 16% of male chemists self-report as single, 30% of female chemists are single. Of those who are married, men are twice as likely to be married to a non-chemist as a woman chemist (59 vs. 30%). Even more interesting, 88% of women have spouses who work for pay, compared with only 63.5% of the men. One of the more telling differences is that 40% of women have dependent children, while 50% of the men do. This finding may be one of the factors that drive the larger number of women choosing part-time employment (Figure 4).

FAMILY STATUS OF ALL CHEMISTS BY SEX, 1990-2000	ALL CHEMISTS			ALL MEN			ALL WOMEN		
	YEAR			YEAR			YEAR		
ALL CHEMISTS	1990	1995	2000	1990	1995	2000	1990	1995	2000
SINGLE, NEVER MARRIED	15.3%	15.1%	13.4%	12.6%	12.6%	11.0%	27.1%	24.1%	20.9%
SINGLE, PREVIOUSLY MARRIED	6.0	6.1	6.2	5.0	5.2	5.0	10.6	9.3	9.7
CURRENTLY MARRIED/PARTNER TO A CHEMIST	10.3	11.5	12.7	8.4	9.3	10.3	18.7	19.4	20.0
CURRENTLY MARRIED/PARTNER TO A SCIENTIST	11.9	13.4	15.5	10.8	12.0	14.3	16.5	18.3	19.2
CURRENTLY MARRIED/PARTNERED TO A NON-CHEMIST	56.5	53.9	52.2	63.1	60.9	59.4	27.2	29.0	30.3
TOTAL	100.0%	100.0%	100.0%	100.0%	100.0%	100.0%	100.0%	100.0%	100.0%
% MARRIED WITH SPOUSE WORKING FOR PAYCHECK	–	–	68.9%	–	–	63.5%	–	–	88.3%
% WITH CHILDREN DEPENDENT	47.8%	46.7%	47.5%	51.0%	49.4%	49.9%	33.9%	37.0%	40.3%
% WITH ELDERS DEPENDENT	–	3.2%	3.7%	–	3.0%	3.4%	–	3.9%	4.7%

Figure 4. Family status of all chemists by gender, 1990–2000[5].

Overall Demographics

Women chemists are typically younger than their male counterparts (Figure 5). The most striking demographic difference is the large decrease of white chemists over the 10-year survey period. The decline from 90 to 84% is somewhat compensated for with a commensurate increase in chemists of other ethnicities. These demographic trends, combined with the NSF data on high school graduates who are not choosing the chemical sciences as a potential field of study, are some of the key items driving the multiplicity of studies discussed in the introduction. Continuing to encourage those of other races, as well as women, to pursue the chemical sciences field is one of the best ways to overcome the gaps predicted by the NSF.

DEMOGRAPHICS OF ALL CHEMISTS BY SEX

	ALL CHEMISTS			ALL MEN			ALL WOMEN		
	YEAR			YEAR			YEAR		
ALL CHEMISTS	1990	1995	2000	1990	1995	2000	1990	1995	2000
AGE									
20–29	11.0%	9.4%	7.0%	8.1%	6.9%	4.7%	24.0%	18.5%	13.7%
30–39	32.4	30.8	27.4	30.8	28.2	24.3	39.6	40.2	36.9
40–49	28.3	27.9	29.8	29.7	29.1	30.3	22.3	23.9	28.4
50–59	19.6	23.1	26.1	21.7	25.7	29.3	10.4	13.8	16.5
60–69	8.6	8.4	9.5	9.7	9.8	11.2	3.7	3.4	4.4
70 OR OLDER	0.0	0.3	0.2	0.0	0.3	0.2	0.0	0.2	0.1
RACE/ETHNICITY									
HISPANIC	1.4%	2.2%	2.7%	1.3%	2.0%	2.3%	2.0%	3.0%	3.7%
NON-HISPANIC:									
WHITE	90.5%	84.7%	83.9%	90.9%	85.4%	84.9%	89.0%	82.3%	80.8%
BLACK	1.2	1.4	1.8	1.0	1.1	1.5	1.9	2.2	2.6
AMERICAN INDIAN	0.3	0.2	0.2	0.3	0.2	0.1	0.5	0.2	0.3
ASIAN	6.0	10.2	10.5	6.0	9.9	10.0	6.2	11.2	11.9
OTHER RACE	0.6	1.3	1.0	0.6	1.4	1.1	0.5	1.2	0.7

NOTE: Hispanics are an exclusive group for Race/Ethnicity. All other groups are non-Hispanic.

Figure 5. Demographics of all chemists by gender[5].

Why This Book, and Why Now?

The ACS' PROGRESS program surveyed a number of women, gathering data on perspectives from women in all aspects of the chemical professions—industry, government, academe, entry-level to experienced, and nontraditional chemistry careers. The information obtained from these women was surprisingly consistent in identifying needs and gaps in the current system. Resoundingly, the surveyed women indicated a need for more role models, and more examples of women who are progressing along career paths as illustrations of how varied the faces of success in chemistry can be.

The WCC undertook the challenge of finding successful women willing to serve as these role models. Through interviews, these women tell their stories of careers in the chemical professions, in a personal and refreshingly honest way. Although not the sole answer to many of the questions posed in the opening, these stories illustrate the varied ways women can succeed in the chemical professions, and illustrate the challenges they face as well as innovative ways of overcoming those challenges, and the joys of being a woman chemist.

References

1. 2002 Census of Women Corporate Officers and Top Earners in the Fortune 500, *Catalyst*, 2002.
2. *Chemical & Engineering News*, September 23, 2002.
3. National Science Board "The Science and Engineering Workforce-Realizing America's Potential", National Science Foundation, August 23, 2003.
4. National Science Board, 2004, *Science & Engineering Indicators 2004*, Volumes 1 and 2, Arlington, VA, National Science Foundation.
5. ACS Chem Census, 2000.

Chapter 2

Susan B. Butts, Director of External Technology: Major Chemicals Manufacturing

Arlene A. Garrison

Office of Research, University of Tennessee, 409 Andy Holt Tower, Knoxville, TN 37996

A traditional career that spans many years can still be achieved at a single company. This profile of a Director at The Dow Chemical Company illustrates the slightly different outlook a woman can bring to a traditional company, particularly when she is in a division that is strongly male dominated, such as research management.

Her long and successful career in the chemical industry started with the children's book *Our Friend the Atom*, in particular the chapter about Marie Curie and her discovery of radium. While Susan Butts had no direct contact with scientists as a child, the book and encouraging teachers led her to believe that it was possible to become a scientist. At the University of Michigan she had the chance to participate in research as an undergraduate. Like many college students, Susan was considering a career in medicine until her honors chemistry professor provided an opportunity for her to participate in a summer research project.

During her undergraduate years Susan worked on a total of three research projects, one resulting in her first technical publication. She then went on to receive her Ph.D. from Northwestern with a specialty in organometallic chemistry. Dow has been her only employer and she currently serves as Director

Susan B. Butts (Courtesy of Jeffrey R. Glenn.)

of External Technology with responsibility for Dow's sponsored research at over 150 universities, institutes and national laboratories worldwide.

Moving to Management

Susan started with Dow immediately after graduate school and spent several years in the laboratory before making the transition to management. Research and Development management positions at Dow are normally filled by individuals with technical backgrounds and experience within the company.

Susan has held several positions that provided exposure to many aspects of the company. One period she considers important to her success is the time she spent in Safety and Regulatory Affairs. This position provided a different perspective and interesting opportunities to interact with many individuals in the company. The variety of positions allowed her to develop skills necessary for external interactions. Throughout her management career, Susan has also maintained a current knowledge of advancements related to her technical background. Expansion of her career has required careful scheduling due to the increasing number of job elements requiring attention.

Balance

Susan has always focused on work–life balance and has made choices through the years to support that balance. She even speaks regularly to women's organizations about work–life balance. Her husband is also employed by Dow and they have two teenage sons. She indicates that scheduling is the key to balancing the many elements of her career and family. A plan is essential to be sure that all aspects receive the appropriate attention. At this stage in her career, Susan indicates that the thorough planning results in a loss of spontaneity, which she hopes to regain as her children leave home and the scheduling becomes less complex.

Scheduling and compartmentalization are the key skills Susan identifies to achieve work–life balance. She believes it is critical to distinguish what you want from what others want for you and prioritize accordingly. Be particularly careful and aware of situations where others may want something *from you* rather than *for you*. Other suggestions for achieving balance include:

- Play to your own strengths and avoid your weaknesses
- Accept things you can't change but seek satisfying solutions
- Avoid destructive competition
- Celebrate your successes (especially small things)

Climate Changes

The work climate has changed significantly during Susan's career. There is now an expectation of much higher productivity. The pervasiveness of communication has resulted in the tendency of work to creep into the personal domain.

She notes that there is little time to reflect on projects in progress. She believes that the lack of reflection results in losses that are difficult to measure but very real.

Success

Susan defines success as the reflection of what a person has done – not the titles or possessions of the individual. She believes successful leaders contribute to the workplace in many ways including providing support to assure that company employees enjoy their jobs. She believes that successful leaders must demonstrate alignment with their personal values and must measure success by considering all facets of life.

Success has required some choices for Susan and she notes that we all make choices daily. Susan chose to delay having children for the first five years in order to establish her career at Dow. Then, while her children were very young, she chose to work less than full time for four years. Fortunately family-friendly policies at Dow allowed Susan to take the time to attend important school events with her sons. She also avoided jobs that required extensive travel while her children were young.

Mentors

Mentors and advisors have been valuable to Susan. Her first mentors provided technical expertise. For example, Susan's graduate thesis advisor served as a mentor as well as an advisor. Susan notes that she selected a thesis advisor after negotiation of expectations. Susan has also had a number of corporate advisors within Dow. Some have been ongoing through her career while others have provided specific insight in a particular situation. She indicates that many types of mentors are valuable and strongly encourages women to establish such relationships. Her most valued mentor is her husband.

Choices and Consequences

Susan warns women to avoid the guilt trap. Successful women must learn to say "no" and to delegate without experiencing guilt. She notes that guilt is often connected to satisfying someone else's priorities rather than your own. Make choices based on your own priorities and stick with them – don't be a victim. Each person must accept the consequences of their choices and guilt is not a useful emotion in the process. She also indicates the importance of avoiding resentment of individuals who have made different choices. There are many paths to success and each person must be comfortable with her own choices.

Susan's primary recommendation is to make your own choices and happily accept the consequences of those choices. She also cautions against spending time on destructive competition. In many cases she has seen that it can be better to fully analyze a situation and make the choice to not compete rather than waste the time and energy associated with competitions that are not aligned with personal or company priorities. She notes that men and women define success differently and often men become engaged in competition when collaboration would better serve the situation.

On a lighter note, Susan took up golf many years ago because her husband loves to play. She laments her large handicap but has chosen not to commit the time necessary to become a "good" golfer. She lives with the consequences and just tries to have fun. Due to her mediocre level of play Susan does not try to use golf as a social opportunity in business. It is just a chance for her whole family to spend some pleasant time together.

Chapter 3

Anne DeMasi, Regulatory Specialist, A Family-Oriented Career Path

Jacqueline Erickson

GlaxoSmithKline, 1500 Littleton Road, Parsippany, NJ 07054

Within the chemical sciences, a career can venture many different paths. This profile of Anne DeMasi shows that it is possible to have a successful family life as well as a successful career. Perhaps sacrifices must be made along the way, but more importantly one must be open to different opportunities in order to best balance work and home life.

Anne DeMasi is currently the Hazard Communication and Regulatory Specialist in the Product Integrity area at Rohm &Haas, where she is responsible for preparing the material safety data sheets (MSDS) and determining the inventory status of all products for regulatory purposes. Preparation of the MSDS involves knowing the regulatory compositions of all the company products as they impact the business and the countries within which certain products can be sold.

Education and Early Career

When Anne first started college at Chestnut Hill College in Philadelphia, Pennsylvania, she planned to attend medical school. However, in college, she met Sister Helen Burke, who became one of her mentors. Sister Burke encouraged her to join the American Chemical Society (ACS) and opened her eyes to a career in chemistry. Undergraduate course work and summer work on National Science Foundation (NSF) programs helped Anne decide to continue

Anne DeMasi (Courtesy of Rohm and Haas.)

her path in chemistry. After graduating from college, Anne spent the summer working in Product Integrity at Rohm & Haas. She then attended Villanova University, where she earned an M.S. degree in Biochemistry.

After earning her M.S. degree, Anne joined Rohm & Haas as a full-time employee in the Analytical Research Department. She initially worked on rheology of polymers, even though rheology was a completely new area for her. She joined Rohm & Haas because she was familiar with the company and believed that it was the type of company within which she could build a career and was also family-oriented.

After five years in Analytical Research, Anne moved into the Industrial Coatings area, where she worked on the product development for wood and plastic substrates. While in this department, she filed two patents and learned a great deal about the business. This area was particularly interesting because she worked with a variety of chemistries and applications. After a few years in this department, an opportunity became available in the Product Integrity area, and Anne switched careers again to move into her current position.

Balance

When asked about sacrifices, Anne replied that she chose to work part-time for several years. Perhaps she made a small sacrifice because the rate of promotion might have been slower, but she is thankful for the opportunity to have had the option to work part-time. She is comfortable with her choice as it has been very beneficial to her family life.

With two daughters, a busy career, and volunteer activities, work–life balance is important to Anne. She has a huge support system and nearby extended family that helps when she travels. Additionally, her husband of 15 years is very supportive and helps as much as possible. Although not a consideration when Anne moved to the Product Integrity area, the ability to work at home when necessary has made it much easier for Anne to balance work and life.

Success

Anne believes that being successful has nothing to do with her job. She says success is a "sense of inner peace" and an "acceptance of what I can and can't do." Success is also about feeling the ultimate level of satisfaction with a healthy career, a healthy body, and a healthy family. Anne doesn't see herself as being driven in one direction or being competitive. For her, success is more about managing multiple goals. She also believes that success does involve compromise because life requires compromise. For those who define success as becoming CEO of a corporation, much more compromise is required, as family and a life outside of work are often sacrificed. Anne believes that you "can't do it all or have it all." All can be differently defined for everyone, and each person must find what works for them to be successful.

Anne has had a very successful and enjoyable career at Rohm & Haas, and she plans to continue with this job. In order to stay current, she attends seminars and job-specific courses. Networking and maintaining contacts in a variety of areas have also been helpful to her career.

Mentors

In becoming successful, Anne's mentors (the "2 Burkes in her life") have been very helpful. First, Sister Helen Burke at Chestnut Hill College was instrumental in Anne becoming a chemist. Then, James (Jim) Burke originally hired Anne at Rohm & Haas and has provided guidance throughout her career. Jim has also encouraged her to participate in ACS activities. Anne joined the ACS while in college and has been involved with the Philadelphia local section since 1991, when she first became a member of the Philadelphia Board of Directors. The ACS has also helped Anne in her career as she gained experience in running meetings and dealing with different groups of people.

Advice

When asked what advice she could provide to others, Anne suggests that it is important to take time to listen actively, gather as much information as possible, and hear what others have to say. Working by consensus and building relationships is also important. Critical skills for advancement include organizational skills, along with a willingness to get involved and make changes.

Chapter 4

L. Shannon Davis, Leader of Process Research and Development: Adventures of an Industrial Chemist

L. Shannon Davis[1] and Amber S. Hinkle[2]

[1]Solutia, Inc., P.O. Box 97, Gonzales, FL 32560
[2]Bayer Material Science, 8500 West Bay Road, MS–18, Baytown, TX 77520

Tracing the path of an ongoing career from the bench to the business world provides insights into the commonality of the skills required to be successful in a variety of career paths. Chemists of all ranks and in many career paths will need good communication skills, good relationship skills, and good common sense to progress.

L. Shannon Davis graduated with a B.S. degree in chemistry from Georgia Southern College in 1984. She went on to earn a Ph.D. in Inorganic Chemistry from the University of Florida in 1988. She was then hired for her catalysis background and began working for Monsanto, at the Pensacola, Florida, site, as a Senior Chemist in the Nylon Intermediates Technology organization. As she describes it, she is a "career Solutia/Monsanto employee; not fitting the expected mold based on today's conventional wisdom on people having 2 or 3 or 4 employers in the first 10 years of their career."

Shannon has held a wide variety of positions in many different locations within the past 16 years. She has gained much valuable experience along the way and is currently the leader of the process research and development (R&D) group for nylon intermediates.

Getting Started

Based on what her parents have told Shannon, she's always had a need to understand the "why" of things since she was a small child. She loved all forms

L. Shannon Davis (Courtesy of L. Shannon Davis.)

science throughout school, biology and chemistry especially. Her fascination with chemistry began in junior high, when she realized the ultimate "why" things worked had to do with atoms and molecules and their interactions. Her high school chemistry teacher really encouraged her interest in experimentation and helped her convert this interest into a regionally award-winning science fair project.

After high school, Shannon was driven to complete her higher education within the timeframe she had set for herself; earning her undergraduate degree in three years and her Ph.D. in four. She thinks that she missed the traditional "fun" undergraduate lifestyle, primarily due to the class load she took, but she doesn't regret it and would not change any of the choices that she made to accomplish these goals. Self-motivation was easier for Shannon because she had a fabulous general chemistry teacher at Georgia Southern, Alex Zozulin, "who was completely 'outside of the box' in terms of teaching style and delivery." He had an incredible talent for reducing very complex items into manageable parts and was so enthusiastic about the fascinations of chemistry. It was contagious, and Shannon was converted in her first quarter of general chemistry! In retrospect, she notes that all of her professors, in one way or another, continued to foster and drive her interest in experimentation and understanding the "whys" of things. For Shannon, the study of chemistry was like peeling back the layers of an onion; as she progressed in her studies, more and more of the physical laws and understandings of how the universe operates became clearer and clearer. She was finally beginning to understand some of the more fundamental "whys" and she continued to be fascinated.

During her last year of college, Shannon opted for a one-year research program with Zozulin as her advisor. Through his tutelage, she says she became addicted to the joys and toils of research. Shannon also notes, in light of the press over the past few years over "math phobia", that she never really enjoyed math. The math classes that were requisite for both her undergraduate and graduate classes she viewed simply as a necessary evil, something she had to endure so that she could learn more chemistry.

Shannon believes she was lucky in that she chose to attend a relatively small regional college that had (and still has) an excellent chemistry department. She enjoyed the advantages of extremely small student-to-teacher ratios and the individual attention of her professors in virtually every class. In fact, Shannon's graduating chemistry class consisted of only seven people; five of whom were women! All of her professors consistently encouraged the students and were very proactive in helping them determine the next steps in their careers.

Graduate school was a special experience for Shannon in that she worked with an incredibly talented group of people and enjoyed a sense of family with her research group and her research director, Russell S. Drago. Also, going to graduate school was a primary goal for Shannon; one that she had planned and strived for since high school and one that she expected to enjoy as thoroughly as one can relish the accomplishment of any long-term goal. Although graduate school was a tremendous amount of work—academically, mentally, and

physically—Shannon earned the degree that she had sought and with a gre
sense of accomplishment, pride, and joy.

The key skills that Shannon learned in graduate school were critical thinkin
and problem solving, tenacity, project management, and independent thinking
These skills have turned out to be some of the most critical ones she has needed
in her industrial career. Shannon relished the detailed study, knowing that these
were the "whys" she had sought, and she has used more of her chemistry
knowledge and know-how than she really expected to in her jobs. This
knowledge formed the foundation for building upon with new subjects like
process chemistry and engineering intricacies, basic business concepts, and the
application of sound technical judgment to new problems.

Career Path

Shannon spent the first six years of her career studying the chemistry and
improving existing processes for the manufacturing of adipic acid and
hexamethylenediamine. During this time, she also became responsible for
overseeing the operation of one of three remaining chemical pilot plant
operations within Monsanto and worked on novel methods for increasing the
value of a number of co-product streams for the Intermediates business unit. She
also met and married her husband, a colleague at Monsanto, while in Pensacola.
In 1994, she was promoted to Manager, Product Technology in the Saflex
business and moved to Springfield, Massachusetts. She was actively involved in
a key new product introduction and new process improvements. After two plus
years in Saflex, she returned to Pensacola as the manager of the R&D group
responsible for the technology in the nylon-6,6 carpet business.

In late 1997, Shannon moved to Solutia's corporate headquarters in St.
Louis, Missouri, as the Technology Director for the Industrial Products business
unit. This specialty chemicals business unit consists of five different businesses,
which are extremely technically oriented and have widely varying technology
bases. In that role, she was responsible for the currently practiced, primarily
proprietary technologies, which cover the fields of heat-transfer fluids, water-
treatment chemicals, aviation fluids, metalworking fluids, and Solutia's only
chiral molecule. Shannon's next job was in commercial development, where she
developed growth plans targeted to develop new platform businesses; one of
those included acquisition strategies for the new pharmaceutical services branch
of Solutia. She moved back to Pensacola in 2000 and into her current position.
Shannon touts experience as her number one asset. Having the opportunity to try
out new things in her job(s) and to learn from them, whether they were successes
or failures.

Work Climate

Shannon says that today she is a little less naïve than when she started graduate school. In her words, "While I was aware of gender issues, and particularly the lack of women in the physical sciences, I can't honestly say that I was a victim of them or that I believed they were 'real'. In an undergraduate class that ended up in my senior year with a makeup of 70% women, I believed that the pipeline theory was working. The evidence that it was surrounded me. My entering graduate class was 25% women. We had between 5 and 10 women in the inorganic division plus 3 female post-docs on our floor during the 4 years I was in graduate school." Shannon never thought during her entire educational career that being female was either an advantage or a disadvantage in terms of opportunities, grants, or possibilities. She was limited only by her own abilities and by what she wanted to do. To their credit, she says her parents and teachers (graduate and undergraduate) were key contributors to that perception due to their generally unbiased approach to learning and life.

Thus, Shannon was very surprised initially when she started in industry and realized that she was the only professional woman in the building and 1 of only 2 in a 70-person department. This was the first of several "consciousness-raising events" for her. She became increasingly aware of the need for role models for women like herself. After her involvement with an "Expanding Your Horizons" workshop for preteen girls, Shannon was more aware of the need to reach girls in their formative years.

After a few years in industry, she says she got so used to being the solitary woman in meetings that she was actually surprised if another woman attended. As she suspects many women have, Shannon has been asked to take meeting minutes and/or make coffee, which she has gracefully refused, generally. She has also had her ideas and suggestions ignored in meetings until someone male brought it up again, ad infinitum. However, she has never personally experienced some of the horror stories that do exist in terms of harassment or discrimination during her career, for which she is very grateful! In fact, Shannon believes that most of the overt discrimination was banished from the workplace long before she arrived. She thinks that most of the challenges she has faced in her career have been generic challenges and not gender-based. Shannon considers herself lucky to NOT be "the first woman to do XXX" over her career and fortunate enough to have followed women who broke the first ground in many areas in her company. She still waits, however, for a woman-led chemical company to be a norm and not a newsworthy event.

Overall, Shannon believes the work climate has definitely changed since she began her industrial career. What began as tolerance, and in a few rare cases genuine hostility, has evolved to a higher level of sensitivity, action, and concern over issues, many of which are not actually gender-specific. These include establishing "norms" for meeting behaviors, flex time, family leave, and acceptable work behaviors, among others. Additionally, identification of role models, formal mentoring and coaching, and recognition that some behaviors are gender-specific, requiring different approaches to effect change, are all things that have changed over the past 10 years. She also sees more of the pipeline women in her company asking the right questions to succeed: What do I need to do to achieve my goals? What skills do I need to acquire? How did you do it?

Work–Life Balance

On a more personal note, Shannon is part of a dual-career couple. Her husband also works for Solutia, in marketing/technical service, and they have had to work through the challenges of multiple moves in both careers. They have been very fortunate to have management that was willing to work with them to accommodate both of their career needs; a trend far more common today than 10 years ago. Shannon's favorite hobbies are growing hybrid tea roses (when she is in a location where the climate permits), fishing, and reading.

The biggest sacrifice for Shannon and her husband as a dual-career couple has been moving. They have moved four times in the last seven years for her career; a significant stress on people and relationships. Some of the impacts of moving as much as they have include having to maintain friendships long-distance (via e-mail, letters, phone) and not having family or their closest friends nearby and convenient. They have also had to repeatedly re-establish themselves in their community—socially, personally, and professionally—which includes civic affairs, doctors, schools, new work relationships, new neighbors, etc. Moving concerns also include shouldering the financial impact of buying and selling houses, cost-of-living differentials, and taxes; while covered by most relocation benefits, these factors are still critical in the decision to move.

In Shannon's case, her husband has had to sacrifice more than she for these moves, as his career has been disrupted each time they've uprooted. The company and their management have been extremely helpful and understanding in accommodating Shannon's husband's career needs, but frequent changes of business units and technology fields have impacted him in his career progress. One factor that has made the frequent moves so successful is that Shannon's husband's skill set and his ultimate career interests can fit into a multitude of business units.

Shannon reminds us that work–life balance requires a great deal of effort. She says her philosophy sounds much easier than it is. "Remain cognizant of your life goals, and make choices commensurate with those goals."

Success

Success is very personal, according to Shannon. She is very goal-oriented; thus, achieving goals is a key measure for her. However, she also defines success as personal happiness, life balance—being part of a loving family, and having great friendships that endure the test of time and distance. These factors all combine to create the ultimate success for Shannon: the ability to do what she has chosen to do and know she has made a difference when she is done.

Shannon notes that success does require sacrifice as "all things have a price." She goes on to say, "Some are big, some small, some so insignificant that you don't notice a price was paid. The key is to know what you want to accomplish, and to set limits on what you're willing to do and the price you are willing to pay to get there. Balance is the result of solving that equation." Shannon also reminds us that each person's and each couple's balance will be different. From a couple's perspective, agreeing on the really important things up front helps to dictate when and where compromise is needed. Life goals change with age, so flexibility and communication with one's partner is also important.

Mentoring

Shannon has had several mentors and coaches during her career. In her case, her principal coaches have thus far all been men; some were her direct supervisors, while some were colleagues from both inside and outside the company. The ways that these mentors have helped Shannon the most include being a sounding board for ideas—how to present them so that co-workers in different fields could understand, buy-in, and support them. Also, these mentors taught the nuances of corporate politics and how to function in that framework and generating a network of contacts inside and outside the company. Finally, she could seek out these mentors when she didn't know how to get something done or needed advice, guidance, or directions. Shannon's mentors have called to volunteer their time to work with her, and she has also called people whom she admires and requested that they work with her. Shannon states that both approaches work, and she encourages others to take full advantage of any mentoring opportunities available.

In general, Shannon believes that the most important aspect of good mentors is that they want to mentor others and are willing to learn from their mentees in return. This also means being willing to devote the time and energy required to the relationship. The desire to coach others to help them achieve more and to share your hard-earned learning is critical.

Other skills help make the relationship work, such as basic coaching skills, listening, constructive feedback, and empathy.

Over the years, Shannon has found herself more and more in the role of coach herself and she has discovered two pet peeves. First, people who think success is an entitlement that doesn't require sacrifice; second, people who don't know what they want and expect you to have easy answers to what they want to do or be and how they need to attain those goals. She notes that it is much easier to coach people who already have done the hard work, including the introspection required to determine their goals and their balance points in terms of sacrifices and gains.

Shannon has also tried hard to share the lessons she has learned with her peer group and with others. One of her most successful endeavors lately was organizing a get-together of a group of women from different careers to chat about work and family balance, career issues, and other topics in a casual, relaxed atmosphere. These conversations have been incredibly rewarding for Shannon and for all of the participants.

Professional Development

Also aiding in her success, Shannon spends approximately five hours each week on professional development activities, mostly outside of normal work hours. These activities include reading books or listening to them on tape, reading journals and articles on innovative technologies and basic competitive intelligence, and researching on the Internet. She notes that "Webinars", or seminars broadcast on the Internet, are great ways to catch up on new developments or to learn something new without high travel costs.

Shannon recommends reading *The Fifth Discipline*, by Peter Senge, a great book on organizational behaviors and the impact nonsystems-oriented thinking has on organizations at large. Her other favorite is Clayton Christensen's *The Innovator's Dilemma*, which is an excellent treatise on innovation and the difficulties inherent in innovating within large corporations and mature product lines. One reference that she has frequently used recently has been *The 10-Day MBA*, by Steven Silbiger, which is a great overview of basic business skills.

Shannon has also made it a habit to update her resumé whenever she changes jobs and every February after her annual performance review. She has found that it is much easier to put together the updates, changes in responsibilities, and significant events, (patents, publications, presentations) if they are relatively fresh, rather than waiting several years to recreate the information from haphazard files and records. She recommends this practice for others as well.

Not only are updated resumés necessary for sudden job opportunities but also for participating in volunteer organizations, on supervisory boards, and the like.

Leadership

Skills critical to being a successful leader, according to Shannon, are curiosity, excellent communication skills, a willingness to stand up for, defend, and share your work with others and being inspirational enough to encourage others to pursue the field. Additionally, she states that if you directly manage people, you also need good listening skills and empathy, combined with the will to make hard or unpopular decisions when necessary. Similarly, the traits leaders should be looking for in strong engineers and chemists for their team are curiosity; willingness to learn new things; doggedness to follow leads through to their conclusions; thoroughness; critical thinking skills; and the initiative to seek feedback and peer review of their experiments, data, and conclusions.

In order to advance as a leader, one should always be willing to learn new skills. Shannon's favorite training class is Situational Leadership, which adapts leadership behaviors to the four general classes of subordinates. It is taught using the classic film "Twelve O'clock High" to illustrate how one's leadership style is dictated by the group being led and how different groups in different stages react to different leadership styles. Success as a leader depends on your ability to accurately gauge the style of your subordinates so that you deliver what they need in an appropriate and usable fashion.

Additionally, as a hiring manager, you have to be aware of current employment trends, expectations of new hires coming into the market, and salaries. You also have to be aware of the attractiveness of other jobs for your existing employees and strive to compete in your total job offerings. For these reasons, Shannon says that she remains fairly in tune to the current job market herself.

Final Advice

Finally, Shannon believes in two critical skills for advancement of technical people: solid technical know-how and communications skills. A solid background in your field enables you to ask questions, discern good from bad hypotheses, and to use the critical thinking and analysis skills inherent in the sciences. Communication skills enable you to present your work and findings in such a way that they are actionable and will produce results.

Additionally, she states that you should include the terribly cliché items of hard work, dedication, preparedness, and a hefty dose of luck and good timing for ultimate success. She notes that, "timing tends to be overlooked but can be very important. Pasteur said that 'chance favors the prepared mind' This is true not only for serendipitous experiments but also for your career. Hard work and dedication early in your career build the foundation and solid reputation you need to reach your goals. In terms of requisite skill sets, professional presentation skills are critical, as are a solid technical background that provides you the confidence to make decisions and the ability to clearly, succinctly communicate the progress of your projects and your personal career goals. These are the items that will prepare you to take advantage of the luck and timing components."

Shannon also reminds us that because we will be spending a lot of time doing it, a career should be made up primarily of things we love to do. "Solving problems, creating new materials, and discovering uses for them, or finding new and improved ways to make a product are all fun, rewarding activities. A career should be intellectually stimulating with opportunities to learn new skills, and it should be rewarding on a variety of levels. Most of all, a career should hold its proper place in the big picture of your personal values space. For some, a career may be a priority item above all else. For others, their personal values may dictate that family and other life issues take precedence."

Chapter 5

Sally Sullivan, Patent Attorney: Combining Chemistry and Law

Ellen A. Keiter

Department of Chemistry, Eastern Illinois University, 600 Lincoln Avenue, Charleston, IL 61920

Since earning her Ph.D., this interviewee has followed a varied career path, beginning with chemical research and eventually leading to partnership in a patent law firm. Her experience demonstrates that professional growth can include branching into a new area and starting a business that provides services to the larger scientific enterprise.

Sally Sullivan believes it was the environment at her all-girl high school as much as the subject itself that led her to choose chemistry as a field to pursue. Acting on that early decision, she enrolled as a chemistry major at Hunter College of the City University of New York where again she found a particularly nurturing environment. At the time, the campus she attended was concentrated in a single 16-story building in the heart of the city. The chemistry department included a small number of majors and several newly hired, energetic young faculty members—a combination that produced a close working relationship between professors and students. Each student had the benefit of very significant attention and personal encouragement from faculty mentors. Sally gained valuable laboratory research experience with an excellent mentor, Professor R. L. Lichter.

To describe how thoroughly she was groomed and prepared in this environment for her next step— graduate school—Sally says she was "wrapped in a box and sent off."

Sally's graduate school destination was the California Institute of Technology (Cal Tech), where she acknowledges she had to adjust to being in a larger world. Nonetheless, she had a good experience in large measure because she chose a research group where she felt comfortable. In fact, her choice was based as much on the character of the group and the personality of her advisor,

Sally Sullivan (Courtesy of Sally Sullivan.)

Professor J. L. Beauchamp, as on the specific research area. One of the characteristics she most appreciated was the culture of teamwork Professor Beauchamp fostered. She was the first female in her group but others subsequently joined, demonstrating that they too found the environment welcoming. Other female graduate students in the department likewise congregated in groups where they felt comfortable, a tendency Sally observes is still prevalent among women, herself included. The comfort element has remained a strong determining factor in her personal career choices.

Research

After four years at Cal Tech, Sally was awarded a Ph.D. in physical organic chemistry and went on to a prestigious postdoctoral appointment in the laboratory of J. M. Lehn at Université Louis Pasteur in Strasbourg, France. She returned to the United States in 1979 and assumed a second postdoctoral position at the University of Colorado with Charles DePuy. Her thinking at the time was that it would be easier to pursue permanent employment from a domestic base. However, by the end of that appointment, she had met her husband and had decided not to leave Boulder. Thus, as she puts it, she "had to find something to do". Not being interested in an academic position, she sought other opportunities.

Branching Out

Thanks to her connection to Cal Tech, Sally heard about Synergen, a new biotechnology company in her area. At this point, another central element in her career path—flexibility—becomes especially apparent. Although biotech was not a familiar field for her, she applied and was hired. Although not totally happy there, she learned some very important things about herself during the five years she spent with Synergen: 1) her scientific instincts were correct, 2) she could write well, and 3) she could conduct thorough library research. These three strengths led Sally to a successful and comfortable niche in writing proposals.

Through another Cal Tech connection, Sally was introduced to a patent counsel at Agrigenetics, a plant molecular biology company in the region. She was subsequently offered a position with that firm in its patent division, which she accepted. As she became involved in writing patents, she came to realize that it was "clearly what she was meant to do." She soon became a registered patent agent, and her career in patent law was officially launched. Sally's experience at Agrigenetics involved a lot of writing and communicating with scientists, two roles in which she felt both very comfortable and very competent, although she isn't sure which came first. She's grateful that she had an in-house position in which to learn the field of patent law because it provided a lower stress environment than, for example, working in a law firm. She also felt very comfortable working with Lorance Greenlee, the person who had hired her.

A major turning point in Sally's career came in 1987 when Agrigenetics was purchased by Lubrizol and moved to Ohio. Her division was invited to relocate along with the rest of the company, but she and two patent attorneys, Lorance Greenlee and Ellen Winner, elected instead to form their own patent law firm specializing in biotechnology and chemistry. Together the three faced the challenge of learning how to set up a firm and acquire clients.

More Education

As another part of this overall transition, Sally soon recognized that she would not be fully independent as a patent agent and decided to enroll in law school. It was clear to her that having a law degree would be good both for the firm and for her position within it. Thus, in 1992 she entered the University of Denver College of Law. Somewhat contrary to her expectations, Sally loved law school, where she learned a great deal and encountered some of her best teachers. Among her fellow students were a significant number who were, like her, older and more experienced, which helped make for a comfortable environment. In 1995, Sally earned her law degree, with honors, and became a full partner in the Greenlee, Winner, and Sullivan law firm.

Adjusting to Change

Sally faced two particular challenges as she adapted to her role as a patent attorney. One was learning to sell herself—something for which none of her formal training had prepared her. Although quite uncomfortable with the process at the beginning, she has gradually found it easier and now approaches it with much greater confidence. A second area that required adjustment was accepting that she is not the science expert in her dealings with scientists. While because of her background, she can ask appropriate questions, it's the scientist clients with whom she's working who are the technological experts; what she contributes is her expertise in law.

Shaping the Work Environment

Greenlee, Winner, and Sullivan is a small, "boutique" firm with about 20 employees, 7 of whom are patent attorneys, most with advanced degrees in science. Although the group recently had an opportunity to merge with a much larger firm, they elected to remain small. A major element in that decision was a wish to preserve the lifestyle and culture of the group. Sally describes the process of seriously considering the merger as very beneficial because it ultimately affirmed the partners' philosophy in establishing and shaping their own independent firm. One indication of the firm's environment is that it has attracted a number of women scientists who want a flexible means of combining family with a profession. Part-time work is allowed, which Sally believes is beneficial not only to individual employees but to the entire firm.

Balance

Like some of her co-workers, Sally has appreciated her company's flexibility, which has enabled her to work while raising a family. She and her husband have two children, the first of whom was born while Sally was at Synergen and led to her establishing the company's maternity leave policy. Their second child was born after the formation of her current firm. Sally doesn't believe that her career choices have required significant sacrifices in this area. Her husband, a professor, has shared in childcare, and they were able to secure good day care when it was needed. The natural routine for the family has always been for everyone to get up in the morning and for each to go off to his or her own activity for the day.

Being involved in her children's activities has been an enjoyable and relaxing way for Sally to relieve the stress of work. She has also enjoyed a variety of craft and volunteer activities, which offer the double benefit of providing an entirely different environment from her workplace and at the same time giving her a sense of productivity. Sally acknowledges that since co-founding her current firm, escaping work responsibilities has become a greater challenge because it's hard not to be constantly on call. As a result, traveling to interesting sites has become an increasingly important leisure activity for her and her husband.

Advice and Reflection

The important elements that have guided Sally's career choices also form the basis for advice she offers to other women chemists: 1) find out what you're good at and enjoy doing; 2) seek an environment in which you feel comfortable, appreciated, and respected; and 3) remain flexible and open to new opportunities. Although she believes women have to work especially hard to be successful, she also believes that gender has been an asset for her in areas such as maintaining flexibility and being sensitive to problems.

If success means liking what you do and knowing that you're doing it well—which is how Sally would define it—she has certainly achieved it. She enjoys her work and feels fortunate to have found a satisfying niche. In looking ahead, she sees challenging personal and professional transitions as her children leave home and she attempts to scale back her work commitments. Important goals for her are to adapt smoothly to these changes and to shift responsibilities within her firm in a manner that ensures its continuation well beyond her time there.

Chapter 6

Louise B. Dunlap, Director of Technology Transfer: A Career in Creating Partnerships

Arlene A. Garrison

Office of Research, University of Tennessee, 409 Andy Holt Tower, Knoxville, TN 37996

A doctoral degree is not essential for women chemists to attain management status. This profile illustrates how one woman leveraged her interests and community expertise to develop a career in technology transfer with a bachelor's degree in chemistry. It also shows the opportunity for interaction between government and industry to achieve common goals for the public good.

Technology Transfer. Economic Development. Technology Partnerships. An interesting career for a B.S. chemist. Lou Dunlap began her career as a bench chemist at the Oak Ridge National Laboratory (ORNL) in the late 1950s. Her vision has grown with ORNL, and she currently serves as Associate Director of the ORNL Technology Transfer and Economic Development Directorate and Director of Technology Transfer. She and her staff work with companies to license technologies developed with government funding, to develop Cooperative Research and Development agreements to expand the uses of government technologies, and with User Facility Agreements to provide access to unique facilities to academic and industrial users.

Personal choices and interests beyond chemistry have heavily influenced Lou's career. She serves on numerous Boards of Directors and received the Woman of the Year in Business and Government Award from the East

Louise B. Dunlap (Courtesy of Louise B. Dunlap.)

Tennessee YWCA in 1994. Her story provides an interesting view into career transitions to early or mid-career women chemists with an interest beyond the bench.

Education and Early Career

As an undergraduate student at the University of Tennessee (UT), Lou worked part-time at ORNL. She received her bachelor's degree from UT and transferred to full-time laboratory work at ORNL. At that time ORNL was a very secretive and focused government organization. Much of the chemical analysis supported the work to develop nuclear devices used by the military. After some years of routine bench work in the chemical and analytical divisions, Lou moved to a position in research. At that time the ORNL facility had little involvement with area businesses or major businesses other than buying materials. The "fence" around the compound was literal and figurative. ORNL emphasized excellent research work for government purposes and was not involved in community job creation. Everything important in Oak Ridge happened "inside the fence."

Community Focus

Lou left the traditional workforce for 10 years and became active in a variety of civic organizations. She was very concerned about the future of Oak Ridge, which had been established for government work, and a need for job growth in the local area was becoming apparent. Lou focused her interest with the Chamber of Commerce, where she worked extensively in economic development and community relations. For long-term regional development, the Chamber emphasized the attraction of related businesses to the area. The departure of the previous Chamber Director created a window of opportunity for Lou, who was named to that position and held it for several years. This position involved working with all the manufacturers and businesses in the Oak Ridge area, including significant interaction with the ORNL management.

Times had changed, and ORNL was attempting to shift focus to a variety of other research interests, including energy. The benefit of an increase in jobs in the area through attraction of new businesses was apparent. In the 80s, ORNL was looking "outside the fence" for many reasons. As a major employer in the area, ORNL saw the need for related businesses in the area to maintain a steady and available work force. Community relationships and business partnerships were becoming a significant part of the laboratory mission. Important metrics for successful management of the lab began to include such items as spin-off companies and external licensing of technologies.

Lou's experience inside and outside ORNL was perfect for the needs of the organization, and she accepted a position with ORNL within the Public Relations office. After some time in Public Relations, she transferred to the Technology Transfer office at ORNL, where she held a variety of positions. Continued strong emphasis on the application of government-developed technology in the private sector led to expansion of the ORNL program, and Lou was named the Associate Director of Technology Transfer and Economic Development. In that role, Lou has helped numerous researchers move their technology from a government laboratory into the community by starting new businesses. Often the start-up businesses are excellent examples of the use of highly complex technologies for everyday applications. The companies also provide economic growth to the area. In some situations the technology transfer organization licenses technologies to large companies. Many of the licensees are major U.S. corporations, and some are smaller businesses. Lou spends much of her time in meetings with corporations looking for the best way to structure a partnership to advance a technology. The licensees must understand the technology, and the relationship must be designed such that the company and the government sponsors who funded the initial work both benefit.

Changes in Work Climate

Lou notes many changes in the work climate over her career, including a recent emphasis on promotion of women in management and on overall workforce diversity. Her experience in a government setting has been very positive. The emphasis on diverse hiring appears to be stronger in government than industry, as she notes the lack of diversity in the corporate representatives who visit ORNL. In particular, the older management representatives of many visiting companies continue to be almost exclusively male.

Another significant change in the work climate is the increased mobility and flexibility of younger employees, which matches well with the increased expectation by management that travel is necessary in the global economy.

Work and Life Balance

On the balance between work and life, Lou suggests that flexibility in hours is the key. While her current position is much less flexible than some she has held, she has significant vacation and no barriers to using the time. She sees volunteer activity as an important part of the balance that provides an additional outlet for talent.

Additionally Lou believes that she has not made any sacrifices for her career. At key points she could scale back work time and participate in school activities with her two daughters, Leigh and Lynn. Her continued involvement in regional organizations allowed her to stay connected and step back into a professional role when home demands declined. She has traveled extensively with her daughters and visits her two grandchildren regularly. Lou mentions travel as a particular interest and finds that her professional career allowed her to explore many interesting places.

Success

Lou defines success as finding a job where you are happy, contributing, and challenged, but not stressed. She thought that her work on the bench was successful when she achieved particular milestones. She believes that men and women often look at success and careers very differently.

Through community activities with the University of Tennessee, area Hospitals and the Oak Ridge Planning Commission, Lou believes that she has had a positive impact beyond chemistry as well. She suggests that women should become involved in local organizations that are aligned with personal interests. These groups can provide a useful network and a means to have a lasting impact. Also the skills developed in a volunteer organization provide a large benefit. Experience and contacts gained from civic work can provide the tools for a new career or for advancement in an active career. It is important to identify an organization with an exciting mission that will inspire the best effort.

Final Thoughts

Final advice from Lou: "Do a good job—and be positive". In her supervisory position she selects new employees and promotes by attitude. She points out that work skills and knowledge can be developed. It is very hard to change an attitude.

Chapter 7

Nancy Jackson, Deputy Director: A Full-Circle Career Path

Ellen A. Keiter

Department of Chemistry, Eastern Illinois University, 600 Lincoln Avenue, Charleston, IL 61920

This interviewee's career path began in political science and progressed through a transition to chemistry, which included 20 years of involvement in scientific research. In her latest transition, she has returned to her original interest in a position that emphasizes policy as well as science.

By the time she was ready to attend college, Nancy Jackson had already gained considerable political experience. She had spent several months working in her senator's office on Capitol Hill, had assisted in the campaign for a lieutenant governor candidate in her home state of Missouri, and had worked for the city of St. Louis. Thus, it was quite natural that when she entered George Washington University she intended to major in political science. Much to her disappointment, her first political science course was not very interesting. Her general chemistry course, on the other hand, was just the opposite. When, at the end of the first term, she received an "A" in chemistry and a "B" in political science, she began to seriously question her original choice. She was reluctant to switch to chemistry as a major, however, because even though she liked the subject, she had never thought of herself as a "technical person".

Nancy credits a long conversation with her general chemistry professor at the end of her freshman year with convincing her that completing a chemistry major could lead to a successful and satisfying career.

Upon graduating from George Washington University with a bachelor's degree in chemistry, Nancy took a position with the American Chemical Society (ACS) in Washington, DC. She spent three years there in a variety of roles but ultimately decided she wanted to do something more technical and, specifically, work on applied problems. That decision led her to enroll in a graduate program in chemical engineering at the University of Texas (UT), Austin.

Nancy Jackson (Courtesy of Nancy Jackson.)

Research

While in graduate school, Nancy discovered that she thoroughly loved research. Although her original goal had been a master's degree, she earned a Ph.D. in chemical engineering from UT and went on to do postdoctoral research in the department while her husband finished his Ph.D. program. At the end of that period, Nancy and her husband both accepted research appointments at Sandia National Laboratories in Albuquerque, NM.

During her 13 years at Sandia, Nancy has been engaged in a number of research endeavors, all of them chemistry-related and most focusing on heterogeneous catalysis. Many of her projects have involved partnerships with scientists in other government laboratories, industries, and universities. Her accomplishments have led to a variety of leadership roles at Sandia, including Chair of the Catalysis Steering Committee and, more recently, Manager of the Chemical and Biological Sensing, Imaging, and Analysis Department.

A New Direction

A few months ago, Nancy's career took a major turn. After two decades of immersion in research, she accepted appointment as Deputy Director of the International Security Center at Sandia. She was attracted to the position because of its combined emphasis on technology and policy, two of her long-standing interests. Although she chose not to pursue a degree in political science as an undergraduate, her interest in policy issues has remained strong and she welcomed the opportunity to make a "full circle" return to that focus.

The center Nancy leads is charged with applying technology in the international security arena. A particular current emphasis is nonproliferation of all types of weapons of mass destruction. Nancy's co-workers include individuals with expertise in a broad range of areas, including physical security. A recent project involved helping the Centers for Disease Control design systems for keeping their pathogens safe which exemplifies their mission.

Mentoring

According to Nancy, she has been influenced by a series of mentors throughout her career and, in her words, they have made "all the difference" to her success. She speaks with particular fondness of her general chemistry professor, whose encouragement set her on the path to a technical career. She recently had a chance to visit with him when she returned to her alma mater to accept a distinguished alumnus award.

As part of that event, her former professor, now retired, took her to lunch and they enjoyed what she described as a wonderful reunion.

Another important person in Nancy's development, although not a mentor in the traditional sense, was Harry Gray, a professor at the California Institute of Technology who she worked with on a committee while she was with the ACS. From observing him, she learned that having an engaging personality can make one a better scientist. She saw that Gray, through his people skills, could attract the best and brightest scientists to his group and allow them to flourish. As a result, his career was enhanced and so were the careers of his co-workers. Gray's example led Nancy to realize that her own people skills could help her succeed in science. She describes arriving at this conclusion as a crucial step in her decision to pursue a technical career.

Nancy has also assumed the role of mentor for a number of colleagues at Sandia who are in the early stages of their careers. Currently she mentors three young women scientists, making a point of having lunch with them at least once a month to talk about career issues. On a less formal basis, she also provides support to a young chemical engineer with whom she shares a Native American heritage.

Based on her experience on both sides of the mentoring relationship, Nancy believes that being an effective mentor requires a level of comfort in one's own career and a demonstrated ability to achieve one's career goals. A willingness to listen and to openly share both successes and failures is also important.

Challenges

One of the major challenges Nancy faced early in her career was a holdover from her undergraduate days. She says that it took her a long time to overcome a lack of confidence in her technical ability and to develop assurance that she could succeed in science.

The hardest challenge Nancy has faced in her professional capacity was dealing with an especially difficult personnel situation. After struggling with the matter for some time, she eventually came to an important conclusion that it was more important for her to remain healthy and sane than to follow a particular career path. She counts this realization as one of several valuable lessons she gained from the experience. Through it she also became aware of the behavior pattern known as workplace bullying. This heightened awareness motivated her to work with Sandia's leadership to put in place strong rules regarding the issue. Another outcome of the experience is that she now recognizes bullying behavior more readily and, when she observes it in any professional setting, she actively defends those who are being bullied and helps them deal with the situation.

Nancy believes special challenges exist for women in technical fields, especially if they are in positions of leadership or hold such aspirations. The underlying cause, she believes, is that women scientists are not perceived as leaders as readily as men. As a consequence, women in scientific disciplines have to work harder to be considered worthy of leadership roles. Although she acknowledges that overt prejudice against women in science has diminished over the past several decades, Nancy believes it still persists in subtle forms. She says "It's not just women being treated negatively but men getting something extra." In her view, the situation is accurately summed up by a conclusion of the Massachusetts Institute of Technology (MIT) Study of the Status of Women Faculty, namely, that "marginalization of women accumulates from a series of repeated instances of disadvantage which compound over an entire … career." Just as the personnel challenge she experienced inspired Nancy to actively work toward elimination of the problem, her response to this challenge likewise has been to look for ways to improve the situation. For change to occur, she believes it is critically important for women in science to vouch for other women as leaders and to actively promote them. For her part, she takes advantage of every available opportunity to support women colleagues and to lend whatever credibility she has toward boosting their careers.

Development

Professional development has been very important to Nancy during her entire career. One avenue she has particularly enjoyed for advancing in her discipline is through technical symposia. As chair of an ACS division and head of a secretariat, she had several opportunities to organize symposia involving leading experts in areas of current interest. She describes these occasions as ideal venues for moving forward and remaining up-to-date technically.

Nancy has also taken advantage of opportunities for professional development offered by Sandia and its managing company Lockheed Martin. For example, she is near to completing an Advanced Sales Training program for Sandia managers. In this program, which consists of a series of periodic two- to three-day sessions over the course of a year, she says she has learned a great deal about determining what customers want and finding ways to satisfy their needs.

Part of remaining current for Nancy is keeping in tune with the job market. This was particularly true during a recent three-year period when she was actively considering a career move. In addition to reading ads for positions in *Chemical & Engineering News* and *Science*, she talked with professional colleagues from a variety of settings to help her gain perspective.

During this period, she applied for several openings, both within and outside of Sandia, and completed the interview process for one external position. Ultimately, she chose to accept the Deputy Directorship because of its ideal match with her own interests and goals.

Balance

When asked how she balances work and family life, Nancy replied, "It's a constant juggle." She and her husband have twin sons who are 10 years old. From the time they were born until very recently, the family employed a woman who served as a nanny for the boys as well as a general housekeeper, an arrangement that went a long way toward easing work/family pressures. It has always been important to Nancy to spend as much time with her family as possible and, at the same time, have a satisfying and enjoyable career. One practice that has helped her achieve that goal has been to travel frequently with her children, which has often meant bringing them to professional meetings. Because her husband is also a chemical engineer, they frequently attend the same conferences, which has made it relatively easy to turn a professional trip into a family outing.

Nancy's general advice to others for achieving a satisfactory balance between work and family life is to recognize their basic values and act on them. A good starting point is to ask oneself what is most important: Is it having the house picked up all the time, or saving all one's money, or doing things perfectly at work? Nancy advises honestly answering questions such as these and then spending time and money accordingly.

Success and Advice

Nancy's definition for success has evolved over the course of her career. At the time she received her Ph.D., she envisioned that success meant recognition within the scientific community as a good researcher in her specialty of catalysis. At this point, two additional elements are important to her. One is being in a position that allows her to use as many of her skills and gifts as possible. The other is knowing that she is making a difference in the world. Not surprisingly, the evolution of Nancy's view of success parallels the progress of her career. The excitement with which she speaks of her current position is clearly rooted in the opportunity it provides her to apply a broad range of skills in matters of significant global importance.

Like her advice for achieving career/family balance, Nancy's basic recommendation to those seeking a successful career is to know what they most value and make choices accordingly. She thinks it is important to step back and ask whether one is merely emulating a particular success model or developing one's own. One of her former co-workers once made a statement she finds quite useful in this context. Based on his experience, he observed that people tend to do what they most like to do. Nancy would urge those planning their careers to apply that idea by asking themselves how they most relish spending their time. A bit of honest reflection can help one identify what will lead to greatest satisfaction. Finally, Nancy recommends defining success as broadly as possible.

Chapter 8

Sharon V. Vercellotti, Company President: A Business of One's Own

Rita S. Majerle

Hamline University, 1536 Hewitt Avenue, St. Paul, MN 55104

A diverse and varied background is essential when starting up and running a small business. This profile details how a woman with a wide variety of research experience obtained the business knowledge and acumen to start her own business in a cross-cutting field.

In the field of complex carbohydrate chemistry and biopolymers, Sharon Vercellotti is a well-established presence. Sharon is the founder, owner and president of V-LABS, INC, an independent, private consulting and manufacturing firm in Covington, Louisiana, established in 1979. They manufacture carbohydrates, polysaccharides, and glycosidic enzymes for research in the food, biochemical, and pharmaceutical industries; as well as in universities and research institutions. V-LABS has utilized membrane separations in the isolation of natural polysaccharides and has developed molecular weight standards of these polysaccharides. Sharon's energy and drive have made V-LABS a successful venture.

Sharon started V-LABS because she saw a special need for biopolymers. Her vision was to make products and services available to researchers in biotechnology, biochemistry, and in the rapidly developing field of glycobiology. Sharon's business has expanded globally and she has made international agreements to distribute biologically active cell surface recognition molecules that are much in demand in frontier research. She has also made available carbohydrate enzymes for use in structural studies and analysis of key biomolecules.

Sharon V. Vercellotti (Courtesy of Sharon V. Vercellotti.)

Education and Professional Experience

Sharon is the daughter of a blacksmith and a nurse and has always enjoyed science in spite of her early experiences. "My beginning chemistry classes weren't very good in retrospect. I almost went into physics," she mused. Fortunately, curiosity and coming from a hands-on background fueled her interest in the sciences and led her into chemistry.

She attended Louisiana State University as an undergraduate, majoring in chemistry. While she was there, Sharon was selected to participate in the National Science Foundation (NSF) sponsored research experience for undergraduates summer program (NSF-REU). She was one of thirty students, selected from all parts of the country, to go to the University of Arkansas for the summer. "It was exciting. There were things to be done! I got to work with graduate students and research professors too. It is a great program and opportunity and I think everyone should take advantage of it," said Sharon enthusiastically. It was this summer research experience that led her to pursue graduate work.

Sharon went on to attend The Ohio State University and obtained her M.S. degree in chemistry in 1965 under former American Chemical Society president Daryle H. Busch. Her research there focused on synthetic transition metal chelation complexes. After finishing graduate school, Sharon undertook a series of research positions both in research institutes and at various universities. Her biochemical research projects have resulted in papers on subjects such as blood proteins, nucleic acid base pair thermodynamics, protease mechanism determination, dehydrogenase purification, and pyridine nucleotide analog synthesis. She has also been successful with active site specific reagents for dehydrogenases based on novel nicotinamide adenine dinucleotide diazonium salts.

During her 16 years as president of V-LABS, Sharon has worked on both the synthesis and analysis of carbohydrates, and the isolation and modification of polysaccharides. She has gained extensive experience in the application of membrane separation to biopolymers. She has also developed membrane purification methods for the isolation of polysaccharides from seaweed and hemicellulose from corncobs. These membranes have also been used by Sharon for the fractionation of maltodextrins and insulin into different molecular weight ranges; which are sold for use as standards in membrane filtration systems.

As a consultant to many food, surgical supply, and pharmaceutical companies, Sharon has solved packaging and shelf-life problems in a wide range of products. She has also developed reliable procedures for scaling up immunomodulating polysaccharides and supplies them to user companies.

Additionally, Sharon has prepared and purified chitin and chitosan from shrimp and crayfish and has prepared several N-carboxymethyl and sulfated N-carboxymethyl derivatives. The latter compound has been shown to be active against HIV-1, the virus that causes AIDS, in human tissue culture. She has worked with the Dana Farber Cancer Institute at Harvard University on this project. She has also prepared a number of sulfated polysaccharides such as xylan sulfates and amylose sulfates as commercial products, useful in therapeutic applications.

Sharon has authored or co-authored over two-dozen publications and has been the recipient of numerous grants and contracts. Iota Sigma Pi, an honorary society for women chemists, recognized her with its 1999 Triennial Award for Professional Excellence. She was also given the honor of Employer of the Year in 1993-94 by Covington High School. Sharon's work in community programs and projects includes participation in economic development and educational outreach. As far as continuing education, Sharon said that the series of small businesses courses she took at Virginia Tech after she had finished graduate school were inspirational and very informative.

Work and Life Balance

In 1978, however, Sharon decided it was time to return to Louisiana. She opened V-LABS in an old building kitty-cornered from her father's blacksmith shop. Here family support was available on all levels, from her parents to her husband, who is the vice president and senior chemist at V-LABS, to her two young children. "The kids grew up in the lab. They were always running around there!" laughed Sharon. Her son eventually became an integral part of V-LABS and installed the computer network system for the company, establishing data bases for lab results, supply ordering, etc. Sharon believes that a balance of work and family life is key to personal satisfaction and success.

Success

Sharon also said, "To be successful is to be challenged every day and learn something new as you are doing it." In addition, she is adamant that networking is essential for success in business. Learning how to secure funding, for her business or many of the symposiums that she has chaired, has been a key to her success. Funding from the NSF-Small Business Innovation Research program (SBIR) has helped make V-LABS what it is today. She in turn has given back to the program as an advisor on a national visiting committee and has served on the program's advisory board.

Mentoring

Mentoring has played a role in her successes too. Sharon credits Dr. Elias Kline, of the University of Kentucky at Louisville, as being a strong mentor. He influenced her development of membrane usage for polysaccharide applications. Also Sharon herself has been a mentor to a number of high school and college students at her laboratory. She has supported internships for the Gifted and Talented Program with the St. Tammany Parish Schools, from 1992 to the present. Many of these students have pursued studies and careers in the sciences.

Final Thoughts

Times have changed since Sharon opened V-LABS, Inc. "I don't get calls asking to speak to the person in charge anymore!" Sharon laughs. She is a vibrant and passionate woman and an advocate for pursuing small business opportunities. "80% of all biotech and pharmaceutical companies are small businesses. In 2002 more graduates were hired into small businesses than large and 58% of all employees work for small businesses."

Her final words of advice were: "Be willing to take risks. Be confident that the call will come. Balance security and risk taking." Advice that has worked well for Sharon Vercellotti.

Chapter 9

Jean Zappia, Vice President of Plastics Additives Segment: Achieving Life Balance

Jody A. Kocsis

Technology Manager, Engine Oils, Lubrizol Corporation,
29400 Lakeland Boulevard, Wickliffe, OH 44092

"Never sacrifice your integrity, work hard, and have fun" is
what Jean Zappia believes in. Jean is the Vice President of the
Plastics Additives Segment of Ciba Specialty Chemicals,
(previously Ciba-Geigy Corporation). Jean tells her story and
provides direction to women entering the sciences.

Jean Zappia has been with Ciba Specialty Chemicals since receiving her
B.S. in Biochemistry from Manhattan College in 1982. After spending 5 years
in Ciba's applications laboratory in Polymer Additives, she moved into product
management, marketing, and sales management positions and spent time in both
the polymer additives and lubricant additives businesses. While working for
Ciba, Jean earned her M.S. in Technology Management from Polytechnic
Institute of New York. She is currently managing Ciba's lubricant additives
business for North and South America and is a member of the NAFTA
Leadership Team for the Plastics Additives Segment (consisting of Ciba's
lubricant additive and polymer additives businesses).

Not only does Jean enjoy the challenge of managing a business and being
ultimately responsible for its profit and loss, but she also enjoys a life outside of
Ciba. Jean and her husband of 18 years, Robert, have two children, Brendan,
age 11, and Emma, age 6. They live in a suburb of New York City, Yorktown
Heights. In her free time Jean enjoys photography, kids' soccer games, home
improvement, and remodeling.

Jean Zappia (Courtesy of Jean Zappia.)

The Decision

Jean always enjoyed science in high school and considered going to medical school after college. While an undergraduate, she began working in a local hospital as an EKG technician to determine whether medicine was the career for her. Although she liked the medical field, she did not necessarily like the "hard exterior" that many doctors had to adopt in order to deal with the human suffering they encountered every day. She decided to stay near medicine, but not as a medical doctor. This decision led Jean on a journey to find a career that satisfied her medical desire but was not in the operating room.

She originally interviewed for a position with Ciba-Geigy because of its reputation as a pharmaceutical company. The company was located very near where she grew up, which allowed her to start her professional career near family and friends. When she learned that she would need a Ph.D. in chemistry before she would be given any real responsibility for running programs, she decided to give the industrial side of the business, in polymer additives, a try instead. Jean found interesting and meaningful work, tremendous people, and the opportunity to move from a technical track (which she realized early on was not for her long-term) to a business career path.

Jean found a company that values its employees, provides opportunities for advancement, and is flexible with regard to the challenges of balancing family and career. She has made some life-long friends at Ciba and has always had fun. The value system she encounters at Ciba is consistent with her own value system: Never sacrifice integrity, work hard, and have fun.

The Work Climate

Jean mentions that over the past few years the business climate has changed dramatically as the chemical industry underwent a protracted period of economic uncertainty. These changes have made it much more difficult to have "fun" at work because the growth opportunities are harder to come by and cost control measures make for painful discussions at times.

Jean has also changed, which she says is a good thing. She believes she is definitely a more mature, steadier individual now than she was a few years ago, which she thinks has helped her endure the rough economic times.

Jean is often asked whether she has faced particular challenges as a woman in the field of chemistry, and in general she says that she has not. She has always thought that Ciba as an organization is fairly gender neutral. In the chemical industry in general, a woman might still be one of the few at any particular gathering. However, Jeans says that she has not faced any particularly tough challenges

Definition of Success, A Compromise

To Jean, happiness equates with success, and her happiness is directly related to how she views her life balance. Is she spending the right amount/quality of time with her family and friends? Does she have time to enjoy pursuits outside of the office? Is she available for her kids when they need her? And on the professional side, is she focused and productive while at work? Does she add value to her company, business, and employees? Has she helped a customer today? When she can positively answer all of these, her life is in balance and she considers herself successful.

Jean also says success does require compromise. When trying to achieve balance in life, one must implicitly prioritize to determine what aspects of life won't be quite as "perfect" as they might be when life is less complicated. So if your house or office is a little less clean or organized, this would be one kind of compromise.

A more important kind of compromise, she thinks, comes from the people around us who are important to achieving success. Her family understands that she has to travel or work late or work at home, and her company understands that she sometimes needs to be with the family for important events. It seems to her that these people are also willing to compromise and in doing so contribute to everyone's success.

Jean believes it is important to listen carefully to the small voice inside that tells when things are getting out of balance. Trust your instincts, and be ready to turn down the promotion if it's not the right time. If you have the right stuff, more opportunities will always arise.

Being both a mother and a professional can be a bit challenging. She had already been a professional for 10 years before having her first child, so she needed a period of redefining the professional woman she had become. What emerged was a professional that perhaps spent a bit less time thinking about work, especially at first, but with a new set of life experiences that she believes has ultimately made her a better professional and manager.

Being a good manager is important, especially in the current climate. Jean thinks managers must challenge their people to help them grow. They need to invest in them, believe in them, and allow them to learn, even if small mistakes are made in the process. Managers need to maintain an open dialogue with their people and continuously inform them of performance—both good and bad. Ultimately, they need to support their people, involve them, and make them aware that their contribution is important to the success of the overall venture.

Career Development Tools

Jean believes in helping people grow. As a manager, Jean has her people partake in a 360-degree evaluation process. She favors 360-degree feedback vehicles, which have many variations and enable employees to view themselves as others see them. She has had great results. She has also participated in a few of these herself and has always found them to be growth experiences.

In addition, Jean has been lucky enough to participate in some excellent management and leadership programs while at Ciba. These courses teach how to manage and motivate teams so that all are on course and achieving the company goals.

Mentoring

Jean was lucky enough to have not one but two mentors in her career at Ciba. Interestingly, they were both bosses, but she met one early in her career and the other much more recently. They were both effective mentors in that they continually challenged her outside her comfort zone, sometimes to the point of putting her in scary or unfamiliar situations, so that she might grow. They had absolute faith in her, even when she had doubts. They talked often (and sometimes argued), but they never stopped challenging her. They were ultimately friends that she trusted implicitly, and they her.

Jean is also a mentor to others. She participates in a Career Day Program at Ciba, where she demonstrates how chemistry is a part of our everyday life. She shows the students that chemicals can be "good", such as plastics and inks. As a mentor, she advises her mentees to not be one of the group: "Be YOU." She also encourages students to take summer jobs or co-op assignments in local companies that they are considering as potential career options.

Final Advice

Career selection is a very personal thing, but for Jean it has always been about challenging work, good people, and the ability to succeed at work and at home. She says, "Don't be afraid to do the things that scare you....ultimately the most personally rewarding growth comes from these experiences. Let the career progression take care of itself....I've seen too many people try to 'play the game' to get ahead....if you do the job to the best of your ability, the promotions will follow."

Chapter 10

Rachael L. Barbour, Senior Chemist: Spectroscopy for Problem-Solving

Arlene A. Garrison

Office of Research, University of Tennessee, 409 Andy Holt Tower, Knoxville, TN 37996

Chemists are trained to use tools and often find their niche in the synthetic or analytical techniques that they know. This profile highlights a spectroscopist who has used her training and abilities to solve a variety of analysis problems for her employers. She finds her value in the broader ability to identify possible chemical problems and design methods to analyze those problems.

"I don't know very many women..." were some of the first words from Rachael Barbour In the interview about her career in the traditionally male petrochemical and concrete additives industries. Rachael entered the workforce immediately after obtaining her bachelor's degree in chemistry at Ohio Dominican University in Columbus, Ohio, in 1979. The series of positions and educational opportunities that led to her current position as Senior Chemist with Degussa Construction Chemicals in Cleveland, Ohio, represents an interesting journey that involved more than the 140 miles that separate the cities. Rachael made extensive use of her mentorships and relationships in professional associations to develop a career using spectroscopy in problem-solving.

Early Career and Education

As a child Rachael initially planned to be a physicist. In college, she changed her studies to biology to meet her need to work with people. In

Rachael L. Barbour (Courtesy of Moto Photo and Portrait Studio.)

chemistry, she found a good mixture of the mathematical sciences and the personal interactions. She comments that one difference to consider between sciences is the number of variables. She was very comfortable with the number of variables in math and physics and found that chemistry involved more variables. However, the variables involved in biological research at that time were overwhelming. Chemistry as a compromise also required less memorization.

Rachael's chemistry degree enabled her to work at the Battelle Memorial Institute in Columbus, Ohio, where she remained for four years. She obtained extensive experience in infrared spectroscopy working with Robert Jakobsen (Jake). Her work was primarily in the environmental area, including air analyses and PCBs. Significant inorganic analysis was involved, which is unusual for infrared spectroscopy. At Battelle, she belonged to a group that was encouraged to think creatively. The group was relatively large, with about 16 professionals, and investigated many different research areas. It collaborated and worked together on solutions. Jake encouraged everyone to become involved in the chemical profession through associations and conference presentations. Rachael was extensively involved in committees on the establishment of standards and analytical methods.

When a position became available with BP Research in nearby Cleveland, Rachael had already established a professional reputation through her work in societies and on standards committees. The move to Cleveland placed Rachael in the role of primary breadwinner, while her husband spent time at home with their two small sons. During her 10 years with BP, Rachael began work on a doctorate in chemistry at Case Western Reserve University. In the BP research laboratory, Rachael worked with a number of analytical techniques, including infrared and Raman in support of products under development. She realized that her true talent was in identifying problems and finding the right techniques to characterize and solve them.

Professional Associations

As a Bachelor's Degree chemist at Battelle, Rachael was strongly encouraged to participate in professional associations and to present technical papers at conferences. She attributes her successful career to early encouragement and a key mentor at Battelle, Jake Jacobsen. He pushed her into volunteer work that helped establish methods for spectroscopic analysis that are still in use. Through committee work she developed relationships that provided access to higher-level positions throughout her career.

The Coblentz society was one of the first organizations where Rachael devoted a great deal of volunteer time and energy. The society is a non-profit organization founded in 1954 to foster the understanding and application of vibrational spectroscopy. Rachael served on many committees and the Board of the society, of which she was elected President in 1991 and was the second female to hold that position. The Coblentz society awards many significant prizes to practicing spectroscopists.

Rachael is also involved extensively in other professional associations and in 2002 was elected President of the Society for Applied Spectroscopy (SAS). SAS was incorporated in 1960 and is best known for its journal, *Applied Spectroscopy*.

In her employment interviews, Rachael has always insisted on continuing her participation in professional associations. The relationships were established in her first job, and she finds tremendous value personally and for her companies in her volunteer involvement. She says that her work with professional associations has provided balance to her life. She also notes that some of her work colleagues do not address this subject in their hiring process and are later discouraged from professional affiliations. She recommends that all potential hiring discussions should be used as an opportunity to clarify the intention to stay involved in your profession.

Preparedness Equals Success

Industrial chemists must be prepared for change, and in 1993 Rachael found herself considering options when BP moved its research facility to England and she faced a layoff. Fortunately at that time, her graduate work at Case Western in physical chemistry was well underway, and her children were in school. Thus, she decided to focus on completing her education and spent two years on additional coursework and research in electrochemistry. She completed the coursework for a Master's degree and was encouraged to proceed directly to the Ph.D. level due to her excellent performance. Unfortunately, health issues within Rachael's family forced her to discontinue her pursuit of the Ph.D. degree and to return to industrial employment.

Because she had remained active in the local professional societies, Rachael located a position with a major chemical additives business in Cleveland, then known as Master Builders. Rachael has remained with the company that is now known as Degussa Construction Chemicals and has advanced to the position of Senior Chemist.

Rachael updates her resumé annually. She does not believe any job is really secure. She monitors the ads in *C&E News* and looks at the local business pages in Cleveland. International market changes and corporate buyouts can happen to anyone in industry, and being prepared is critical.

Rachael sees many different ways to be successful in a chemical career but believes the primary goal should always be to move the science forward. She also indicates that scientific advancement can be found in industrial laboratories as well as academic laboratories and that a Ph.D. is not necessary. She provides an example of the analysis of a critical corrosion inhibitor in concrete. The appropriate concentration is very important in bridges and other structures where concrete is poured around metal. When Rachael began to study the problem everyone agreed that the analysis was impossible in the concrete matrix due to the interferences and low concentration. Two years of work resulted in a reliable in situ method that has improved public safety. Rachael knows this project is a success that pushed back the frontiers of science in a very practical application. Success can also be found in steady, reliable analytical work, particularly in the environmental or regulatory fields. Attention to detail and consistent use of a defined method is essential for long-term interpretation of data, according to Rachael.

Although Rachael does believe that success can require compromise for reasons such as budget constraints, she insists that success never involves a sellout. Working for manufacturers and often on new products, she is aware of situations where salesmen have asked the laboratory scientists to supply reports that emphasize the good qualities of the new product. Sophisticated analytical data are open to interpretation, and unscrupulous management could ask for specific answers. Rachael believes the chemist must maintain integrity for herself and for the science. She has never been asked to lie and would resign if asked to provide an untrue analysis. However, she is aware of situations in other companies where chemists have been forced to make decisions to uphold the truth of their analyses, and she is thankful that she has worked for ethical companies where truth has not been an issue.

Rachael believes that hard work is the key to a successful career, and continues to work long hours. Her work ethic was established while at Battelle, and she has always enjoyed her work immensely. When she moved to industry, she noted two groups of employees: the 9–5 clock-watchers and the folks who stayed when necessary to complete important tasks. Chemical analysis often results in varying demands for time, particularly if samples are time-sensitive. Manufacturing materials may need to be certified before they can be released, or plant problems may necessitate focused work to bring production back on line. Rachael notes that the employees who have shown flexibility and a willingness to work extra hours when required have advanced in their careers much more rapidly. She also indicates the importance of regular personal reflection to identify how one's work fits into the big picture for the company.

Career Resources and Interests

Rachael finds her work fascinating, and much of her pleasure reading is related to her chosen field. Her favorite book is the Autobiography of W. W. Coblentz. The book reflects on his early intellectual pursuits and also describes rural Ohio life in the latter part of the 19th century. Rachael finds the recounting of his early tinkering fascinating. Additionally in her current job, Rachael is encouraged to continue professional development. She tries to spend some time every week, if not every day, keeping up with current research in areas of interest to herself and her company.

The Bowdoin Infrared Course is one of the tools Rachael recommends highly to anyone interested in a spectroscopy career. She finds the instructors Richard C. Lord, Harry Willis, Foil Miller, Dana Mayo, Dick Hannah, and Lionel Bellamy inspirational as well as informative. She notes that good short courses can also provide networking opportunities and can lead to long-term professional friendships.

Family and Balance

Today, Rachael remains married and has two sons, Sam and John, who are currently in college. When the kids were growing up, her favorite pastime was doing anything with them. They both played musical instruments, and music has always been part of their lives. Additionally family vacations often included the beach or camping. In some cases, Rachael had the opportunity to include family in her travel to professional conferences as well.

Unfortunately, her career did require that she sacrifice some of this time with her children. She notes that she was always the major breadwinner and that a reduced time schedule was never an option. At one point she and her husband arranged their shifts so that one of them was home with the children before school and the other after school.

Final Advice

Finally Rachael notes that many young chemists do not participate in professional groups, and she encourages everyone to stay involved with the available associations. Continued participation in professional societies, such as SAS, ACS, and ACI (American Concrete Institute), includes local section meetings and involvement in the national organizations. Advancement of science can only take place if everyone works together and shares information. The associations are also extremely valuable resources in the case of job changes. Rachael believes that her active participation in professional societies has been key to her many positive job changes.

Chapter 11

Elizabeth Yamashita, Group Director of Global Regulatory Sciences: The Challenge of Evolving Regulations

Jacqueline Erickson

GlaxoSmithKline, 1500 Littleton Road, Parsippany, NJ 07054

Regulatory Affairs in the pharmaceutical industry is a growing field with many opportunities for scientists who wish to use their knowledge and skills in this area. This profile highlights Liz Yamashita, a Regulatory Professional in the pharmaceutical industry.

The pharmaceutical industry is filled with exciting careers within the chemical sciences. Some are more traditional laboratory careers, while others are outside of the laboratory. Elizabeth Yamashita started her career in the laboratory but has become successful in a regulatory affairs position. She is currently a Group Director, Global Regulatory Sciences-CMC (Chemistry Manufacturing & Controls), at Bristol-Myers Squibb Company, one of the world's largest pharmaceutical companies.

Wishing to follow in her father's footsteps, Liz became a chemist. She earned her B.S. degree in Chemistry at the University of Rochester and joined Bristol-Myers Squibb after college as a process chemist. Her primary reasons for choosing the company included its location and proximity to the University of Rochester, as well as the friendly people working there. While working process chemistry, Liz was fortunate enough to have a manager recognize her additional talents. Together they worked to find other opportunities within the company where those talents could be better utilized. After nine years in Process Chemistry, Liz switched careers and moved into the CMC Worldwide

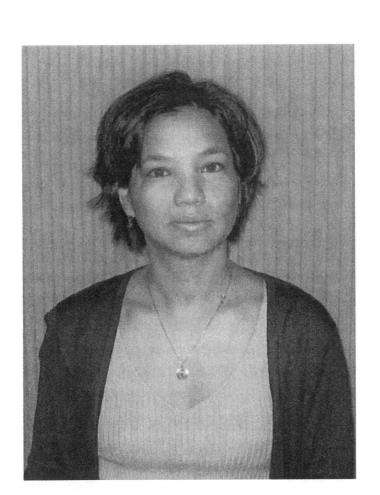

Elizabeth Yamashita (Courtesy of Elizabeth Yamashita.)

Regulatory Affairs Department. A few years later, she was offered an opportunity in the Global Marketing Department, where she performed regulatory-related work. She found this position to be very rewarding, and the experience in this department gave her a perspective on the pharmaceutical industry that not many chemists have the opportunity to obtain.

Balance

After a few years in Global Marketing, Liz transferred back to the CMC Regulatory group, looking for more balance in her life and time to spend with her husband, four children, and pets. In her spare time, she also enjoys cooking and gardening. Liz indicated that she tries to allocate her time in order to separate her work and home life. At home she concentrates on her family, while at work she prioritizes so that she can stay focused and efficient for the tasks at hand.

The Regulatory Field

Regulatory affairs involves the interface between a pharmaceutical company and a health agency, such as the Food and Drug Administration (FDA). Prior to registration or approval of a new pharmaceutical compound, data are reviewed to ensure the quality, safety, and efficacy of that compound. The Bristol-Myers Squibb CMC area focuses on the quality of the active ingredient and drug product. Liz has responsibility for global CMC regulatory strategy, supporting three different therapeutic areas, and has gained experience in every stage of development, from early development to maintenance of marketed products. Currently, she leads a team of approximately 12 people who are responsible for all aspects of CMC regulatory documentation. When asked what she loves most about her job, Liz said that she enjoys the challenges and constant learning that go hand in hand with changing and evolving regulations. She also said that she works with wonderful people who really care about quality work and ensuring that the job is well done.

When asked about professional development, Liz replied that working in Regulatory itself is a form of professional development, as the regulatory environment is constantly changing. In order to work in this field, a person must keep up with new and changing regulations and guidances. Additionally, Liz attends seminars to learn about the latest regulatory approaches or advances in a particular therapeutic area. She networks and stays on top of the job market, both to potentially aid her career and to look for new talent. Liz maintains her Regulatory Affairs Certification from the Regulatory Affairs Professionals Society (RAPS) and is an active member of that organization.

She believes that women who are open and creative in their approach to work would find regulatory affairs rewarding. Regardless of what one chooses, Liz said that work should be fun and personally rewarding.

Success

Liz believes that she has been exceptionally lucky, as she has had managers and mentors who recognized her talents and pushed her to accept challenging projects and assignments. She continues to have excellent managers who push her to try new things as well as provide her with support. Liz noted that two of her three managers in the CMC group were women who understood the need for work–life balance and that these women have provided exceptional guidance and support. In addition to the mentors that she has at work, a great deal of credit goes to her parents who she still relies on for advice. She also believes that success is internally motivated and that "it's about setting your own goals and working to achieve those goals. Success is about making positive choices, and you should never compromise yourself to achieve those goals." She stated that she never had to sacrifice in order to become successful.

Liz also noted that she has had three different careers within the one company, which has helped her reach her current level of success. Although the careers seem very different, Liz indicates that similar skills are used in every position and that all of the positions have had a strong basis in science. The positions all required good communication skills, as do most scientific jobs. Additionally, both Regulatory and Marketing value the ability to anticipate changes, negotiate, and interpret requirements.

Climate Changes

When asked how the work climate has changed over the years, Liz replied that in the past, companies, even large ones, had more of a family feel. In contrast, today's environment expects employees to actively manage their own careers. In many respects this attitude gives the employee more autonomy to go after what they want rather than waiting for advancement to be handed to them. However, movement across companies is often the result.

Final Advice

As advice to others, Liz thinks that you should constantly look for opportunities and "never sell yourself short." Networking is the key to finding those opportunities, and professionals should maintain contacts both within and outside their discipline. She also noted that women tend to take personal responsibility for everything and then become too busy to network, which is a shame. In order to advance in a career, communication and problem-solving skills are as essential as execution or doing a job well. She strongly believes in teamwork.

Chapter 12

Lora Rand, Vice-President of Plastics Manufacturing: Inspiring Leadership

Amber S. Hinkle

Bayer Material Science, 8500 West Bay Road, MS-18, Baytown, TX 77520

Aggressive innovation, extraordinary customer focus, valued employees, and teamwork comprise the vision of the Bayer plastics manufacturing team in Baytown, TX; inspired by their leader and Vice President, Dr. Lora L. Rand. Employees recognize Lora for her commitment to the business and her willingness to take risks. Even so, Lora says you can't take yourself or the work too seriously. Her best advice is to enjoy what you do and make a point of taking time off to rejuvenate.

Lora Rand is currently responsible for 210 employees who manufacture polycarbonate, one of the highest profit margin products for Bayer Material Sciences today. She earned a B.S. degree in Chemistry from Emory University in 1984, studied chemical engineering for two years, and finished with a Ph.D. in chemistry from the University of Texas in 1990. Later that year, she began her career with Bayer as a Senior Process Chemist for polycarbonate manufacturing. Lora's career path led her through quality assurance, laboratory work, and manufacturing, both in Texas and overseas in Germany.

Not long after she assumed her current position, Bayer initiated massive cost-cutting and re-organizational efforts. Thus, as part of the site senior leadership team, Lora has also played a key role in shaping the future of the Baytown site, which consists of approximately 1000 Bayer employees, plus several hundred contractors, and is one of the largest Bayer facilities.

"Lora has been very successful in her current role as a member of the senior leadership team because she is able to communicate a clear vision of expectations. She empowers her teams to focus on the achievement of results, not performance of activities. Lora also ensures that people who deliver are

Lora Rand (Courtesy of Lora Rand.)

recognized for their contributions. Those are signs of a good leader," says John M. Rocco, Vice President and General Site Manager, for Bayer's Baytown site.

Success

When asked to define success, Lora said, "It's not about money or position. For me it's about feeling satisfied. Only if you allow money or position to count towards success do you have to compromise." Integrity, flexibility, consistent hard work, and good networking are some of the factors that Lora believes attributed to her own success. She also places great importance on communication. You must be able to communicate what you want and where you hope to go in order to get there. According to Lora, the ability to communicate with all types of people and within all levels of an organization is critical.

Lora also notes that active listening is of prime importance as a leader. She states that leaders should lead by example and willingly go to battle for their employees. Leaders should also be able to adapt well by being open to change and not blind to their own weaknesses. Lora challenges everyone to take risks, whether they are currently a leader or developing into one.

Lora encourages her employees at all levels to develop their leadership traits. She expects good chemists and engineers to be curious, to work well in teams, to get involved and to be willing to pick up the ball and run with it. Lora especially emphasizes "follow-through" as important to personal advancement. She states, "Find an avenue that makes you happy and pursue it—work at it. Don't just say 'give me stuff to do'."

In fact, Lora recommends that each of us ask ourselves what would entice us to get out of bed each morning and where do we think we can make an improvement. These pursuits make for a satisfying career. She believes it is most important to be happy personally and not to feel stifled in a career. However, she also reminds us that, realistically, we won't be happy one-hundred percent of the time.

As part of her successful development, Lora tries to spend three to four hours a week reading business journals or books. Her favorite books so far include _Good to Great_ by Jim Collins and _First Break all the Rules_ by Marcus Buckingham and Curt Coffman. She touts taking time to read as an important developmental tool and applies any wisdom gleaned to herself and her organization. In fact, she has inspired a formal reading group that meets once a month to discuss such books over lunch.

Mentoring

Lora emphasizes mentoring as a key to success as well. Her personal philosophy is to look to everyone around her as mentors. Not only has she had several formal and informal mentors along the way, but she also learns a lot from her colleagues and subordinates day to day. As a mentor and a leader, it is important to remember that mentoring is most valuable when it is a two-way street, says Lora. Good mentors need to listen as well as share their experiences. Lora also reminds us that in order to find a mentor, you just need to look around.

We should all be learning from what we see people around us doing—both good and poor performers. In fact, some of the best insight comes from learning what not to do by witnessing others' mistakes.

Work Climate

One thing Lora has learned over the years is patience, although she says the work climate has become more flexible and slightly more tolerant toward women advancing. Although the chemical industry is a conservative business, Lora believes being a woman is not detrimental. Instead she believes that being a woman can often help you get your foot in the door because of some intrigue still left in being a woman. But she warns that once you're in, much more is expected of you than "normal".

Lora speaks of a challenge for everyone in that the workplace is becoming more and more fast-paced. With cell phones and e-mail, you just can't get away from the office. Lora says, "Don't over-commit at work or try to work as many hours as possible. Not only is it not effective, but it puts undue pressure on your employees to do the same. This is not a good management technique." Lora believes that we must each make work and life balance for ourselves. When she isn't submersed in the global strategy of manufacturing plastic, Lora can be found sailing into the Gulf of Mexico on her boat, "Totally Mellow".

Leading Through the Fog

As an inspiring leader, Lora has led her team through many chaotic situations over recent years. Her philosophy is to focus on what you and your team can actually control and manage yourselves. Be confident and don't worry about what other people can or will do during the chaos. As an example, Lora has rallied her manufacturing team around some "big, hairy, audacious goals" for quality improvement, in a time when the economy and the business are somewhat unstable. These are goals that Lora's team can directly influence, it keeps them positively focused and they are now poised for the future when the market expands again.

Lora's final advice to others is "know when to laugh at yourself." She explains that by not taking yourself too seriously, you will be able and willing to take more risks. Taking risks is necessary for building your own self-confidence and for challenging the process for your employees.

Chapter 13

Barbara J. Slatt, Manager of Corporate R&D and Corporate External Relations: A Broad Spectrum of Opportunities

Jacqueline Erickson[1] and Elizabeth A. Piocos[2]

[1]GlaxoSmithKline, 1500 Littleton Road, Parsippany, NJ 07054
[2]Clairol Research and Development, Procter & Gamble Company, 2 Blachley Road, Stamford, CT 06922

Barbara J. Slatt is an excellent example of a Ph.D. Analytical Chemist who did not follow the typical career path. Instead, she has used her skills, knowledge, and interests in pursuing a wide variety of assignments at the Procter & Gamble Company (P&G), where she has become a very successful manager.

Barbara J. Slatt is Manager of Corporate R&D and Corporate External Relations at the Procter & Gamble Company. In this role, Barb is responsible for Global Product Safety and Regulatory Affairs, Corporate Biotechnology, and Technical Issue Management. In a prior assignment, she worked as Director of New Drug Development, Commercialization, and Global Market Support at P&G Pharmaceuticals, where she was involved in the area of Women's Healthcare.

When asked how she became interested in science, Barb stated that her mother was a biology teacher and that her dad was the proverbial "rocket scientist" at NASA's Lewis Research Center where he worked on the development of solid rocket fuels and solar cells to power early spacecraft. She says, "They stimulated my interest in science and chemistry at an early age. As a youngster in grade school, I had a chemistry set, which I used to create magic shows for the neighborhood kids."

Barbara J. Slatt (Courtesy of Nancy Armstrong.)

Education and Early Career

Barb further developed her love of chemistry in high school by participating in many local, regional, and national science fairs, where she received quite a few accolades, including the Ford Future Scientists of America and the prestigious Westinghouse Science Talent Search. She states, "I enjoyed a lot of support and encouragement from my high school chemistry teacher and my dad who invested his time and effort coaching me in chemistry and math."

After college, Barb received her M.S. and Ph.D. degrees in Analytical Chemistry from the University of Illinois. While at Illinois, she was a member of Phi Lambda Upsilon, an Honorary Chemistry Society. She also earned an M.B.A. degree from Xavier University and was a member of the American Chemical Society and the Air Pollution Control Association.

Barb joined P&G as a Corporate Staff Scientist in the Environmental Safety Division in 1975. Less than three years later, she took on a special assignment on loan to the U.S. Department of Commerce (Washington, D.C.) as Policy Analyst for the President's Domestic Policy Review of Industrial Innovation. After this six-month stint, she returned to P&G and was promoted to Section Head of Human Safety for the Personal Cleansing and Household Cleaning Products businesses. She then moved to Beauty Care Product Development with an initial assignment to develop a two-in-one shampoo plus conditioner, which later debuted as the brand Pert Plus.

From there, she was promoted to Associate Director, then to Director, leading a number of project initiatives in the Over The Counter (OTC) Health Care Division and, finally, in P&G Pharmaceuticals. Something telling about Barb's success is her reply to the question of what she likes best about working at P&G. She says that P&G "doesn't pigeonhole people into narrow jobs, but offers a broad spectrum of positions to grow into or branch out and acquire new skills."

Success

Barb defines being successful as having fun, being intellectually stimulated, making a positive difference in the world, gaining personal satisfaction from achieving results, being good at what you do, and then getting paid very well for this besides. In order to become successful, Barb was asked if she had to make compromises or sacrifice anything in order to reach her goals. In terms of sacrifice, she replied, "Because I have loved my work, found it challenging and at times exhilarating, and found the opportunities to lead limitless, I put many discretionary hours into it. This meant that I sacrificed developing other hobbies or sports. Also, the choice of a career path often entails an opportunity cost— i.e., by choosing to become a chemist, I gave up other fields which I had the talent, interest, and skill for (e.g., trial lawyer)." She also states that she had to give up the strong desire to "always do it my way," learn to appreciate divergent points-of-view, and gain the commitment of others.

Barb indicated that learning to deal upward was also a challenge, as she was part of the first wave of women with Ph.D. degrees moving into industry and most of her peers and managers have been men. Part of the challenge is the misconception that women are not as technically competent as men. In order to help bridge the credibility gap and obtain a challenging position, Barb suggests that women obtain advanced degrees. Once you have a job, you can then focus your energies on demonstrating your technical competence.

Barb indicated that leadership is important for advancement in any area. Good leaders deliver results that build the business. They should be able to execute tasks well, have exceptional project management skills, and effective communication skills. In addition, they should serve as role models, demonstrating inclusive behaviors and taking advantage of diversity. She also mentioned that many of the skills important in a good manager or scientific leader are the same, but each with a different focus. Managers generally need to have a stronger focus on people and project management, whereas scientists need a greater mastery of their particular technical area. Managers should develop a vision and include others within it, as well as collaborate across organizational boundaries. They should be able to recognize good ideas and put their support and resources behind them. They should use the breadth of their knowledge to integrate a variety of technical options to meet business needs. However, scientists collaborate with their peers to develop and execute a technical strategy. They use the depth of their knowledge to influence the outcome of a project.

Business Climate

Since Barb has worked at P&G for many years, she was asked about changes in the work climate since she started her career. Personally, she found that she needed to adjust her focus as she rose hierarchically in the organization, because "my role now operates from a perspective of 20,000 feet rather than on the ground. I've become less involved in details and more on the big picture." Her role has shifted from "doing" to "leading and enabling". Part of her current role involves placing employees in remote locations "plugged into the work of their colleagues at other technical centers around the globe."

She also replied that the way work gets done has changed since she entered the company. It's now much more collaborative and global. It focuses more on achieving goals that are revolutionary versus evolutionary. More work is accomplished in teams, and organizations are more matrixed. Over the past few years, since P&G appointed a new CEO, the business has improved significantly. The company has a renewed focus and clear strategic direction, and the businesses are better aligned with streamlined work processes and quicker decision making.

Barb has also seen a renewed commitment to diversity over the past several years. One example is the Senior R&D Women's Leadership Team, of which Barb is a member. This group helps identify and nurture future technical and managerial leaders among the women in R&D. The Senior Women support the junior women by providing mentoring, networking, training, and job growth opportunities. As more women (and minorities) move up through the ranks, the mood of the company has become more "upbeat" and "optimistic" about the inclusive culture that has emerged within the Company.

Balance

Barb was also asked about work–life balance and she replied "I achieve balance by working very hard and then playing very hard. In my 28-plus-year career, I have always put my vacations to good use and didn't use that time to simply do work in another locale. I thrived on those vacations as much as I thrived on work. I've found if you like what you do, it's easy to be lured into spending more of your discretionary time on work than the balance you desire requires. Thus, I block out time for the things important to me—family, health, recreation, relaxation, service to others." Her favorite pastimes include downhill skiing, reading mystery or espionage novels, supporting a local animal shelter, and providing homes for several rescued felines.

Mentoring

Barb has had several informal mentors during her career. These supporters believed in her abilities, encouraged her to set lofty goals, and pointed her toward opportunities; such as specific schools to attend, job assignments to take, and contests to enter. They also spread the word about her capabilities and contributions, opened new doors, and gave her straight feedback and advice.

Barb believes that mentoring is very important, which is why she manages the Senior Women in R&D Mentoring program at P&G. In her experience, traits for good mentors include the ability to develop a trusting relationship with that person, availability, and willingness to spend time on the relationship. Mentors should be approachable, people-oriented, empathic, and comfortable working with women. They should also be knowledgeable about the company culture, and perhaps have the same function as the mentee, but not be a direct line manager, so that they can give personal guidance and feedback. The mentor should also want to help develop others. Barb thinks that it's best when the mentor and mentee already know each other, as the mentor is then familiar with the mentee's strengths and areas for growth. Ideally, the pair will have much in common. For example, a Ph.D. chemist may want a mentor who is also a Ph.D. or a working mother may want to be paired with someone who understands her extra responsibilities.

When looking for mentors, a student or young woman should look for someone who can be a sounding board or career coach. Someone who can give confidential advice as well as provide a different perspective on the job or the organization is invaluable. Barb indicates that most women who have moved up in an organization are extremely generous and willing to work with others. Young women should find someone they admire and reach out to those senior women (and men) when seeking a mentor.

Professional Development

Many other avenues, in addition to mentorship, lead to professional development. As her favorite developmental tool, Barb chose the Triad, where an individual meets with her boss and her boss' boss as a way to ensure that the employees' personal and professional needs are being met. Prior to this meeting, the three individuals fill out a questionnaire relating to the employee's skills, as well as their current role and work environment. The various responses from this questionnaire are then discussed to better understand the individual's feelings about her current assignment and career development.

In her current role, Barb often must deal with changes in the external and internal environment, as well as help others to embrace and deal with change. She strongly recommended a course taught by a company called VisionQuest titled "The New Reality—How to Make Change Your Competitive Advantage". For those interested in becoming better leaders, Barb recommended a small book entitled *The Leader's Compass* by Ed Ruggero and Denis Haley. It guides you through the process of defining and communicating your own personal philosophy of leadership.

Final Thoughts

Barb imparts this final bit of advice: "Love what you do. Be courageous, and stand up for what you think. Take reasonable risks. Set high goals for yourself and others, and seize opportunities to solve problems and create opportunities. Embrace change, and adapt to it quickly. Take advantage of unique, once-in-a-lifetime opportunities (in my case, it was my assignment on loan to the U.S. Department of Commerce). Meet every commitment, do more than expected, and develop a sense of immediacy. Learn to anticipate questions, issues, and problems. Develop the skills to communicate well verbally and in writing. Work well with others; you can accomplish so much more by teaming with others different from yourself."

Chapter 14

Cheryl A. Martin, Director of Financial Planning: From the Laboratory to Wallstreet

Amber S. Hinkle

Bayer Material Science, 8500 West Bay Road, MS–18, Baytown, TX 77520

When making career choices, it is important to remember that you are not locked into the laboratory just because you have a chemistry degree. This profile highlights one career that has gone from the laboratory, through marketing and investor relations, to financial planning. Gaining broad experience within a company can be key to success.

Cheryl Martin graduated with a B.S. degree in chemistry from the College of the Holy Cross in 1984. She then went on to earn her Ph.D. in organic chemistry at Massachusetts Institute of Technology, where she studied under Professor Sharpless. In December of 1998, Cheryl joined Rohm and Haas Company (Rohm and Haas) as a senior scientist in the Exploratory Plastics and Plastics Additives area. She has gained broad experience in the ensuing years as she moved into management for Rohm and Haas, first in quality, then research, then marketing, and most recently in the finance area of the company. She was Director of Investor Relations from February of 2000 to September of 2003, when she began her current assignment as Director, Financial Planning. This broad experience has contributed to Cheryl's success and her great enjoyment of her career.

Education and Early Career

Cheryl has always been interested in science and was greatly encouraged in that pursuit by her parents. She was fortunate enough have taken two years of chemistry in high school in an interesting program that enabled the students to perform practical, hands-on experiments. Cheryl also performed undergraduate research and worked as a laboratory teaching assistant in graduate school.

© 2005 American Chemical Society

87

Cheryl A. Martin (Courtesy of Rohm and Haas.)

An interest in the interface between what is done in the laboratory and how final products are used led Cheryl to choose a job in industry and influenced her choice of positions along her career path. Cheryl said, "Even in graduate school I was interested in the uses of what we were working on in the 'real world'". She also admits to auditing business courses while in graduate school and that she "talks too much to work on the bench." Her recent positions at Rohm and Haas, in marketing, investor relations, and now financial planning, have been ideal for Cheryl because she has been working in the interface between the laboratory and the final products. However she still keeps in touch with the chemistry through all of her contacts in the Research Division at Rohm and Haas, as well as participation in the American Chemical Society (ACS).

Work Climate

Since she first started with Rohm and Haas 15 years ago, Cheryl believes she has changed appreciably in many ways, including gaining a broader perspective on how what is done at industrial companies fits into the overall global economy and marketplace. Having worked in technology, marketing and finance, she understands how trends like globalization, information technology and regulation have impacted the chemical industry and what companies can do to position themselves to succeed.

This broad experience has given Cheryl a much more strategic view on how to promote success for herself and those around her as well, which is very important since today's work climate promises less job security. Cheryl explains that this means professionals should constantly evaluate what they want, what they do well, and what they can do to further their marketability. In this vein, she states that tapping into personal networks is more important than ever.

Career Development

When she began moving into the business side of Rohm and Haas, Cheryl found that her Ph.D. in Chemistry had not quite prepared her for the change in venue but that obtaining a full M.B.A. was not necessary either. She has taken some accounting courses and highly recommends taking short, focused courses on practical subjects such as negotiating or project financing. In fact, Cheryl says it is important for research scientists to learn what factors influence funding of a pet project. However, her favorite course was on selling and how to understand the unconscious methods of engaging someone to better understand their motivations, which was a course that she took early on in her career.

Cheryl estimates that she spends approximately 5% of her time on professional development, including the experience she gains from volunteering outside of Rohm and Haas. She also remains aware of the outside environment, the general state of the chemical industry, and trends in the job market. She updates her resumé once per year or whenever she takes on new responsibilities.

Success

Cheryl defines success with her favorite quote from Ralph Waldo Emerson, "The high prize of life, the crowning glory of a man is to be born with a bias to some pursuit which finds him in employment and happiness—whether it be to make baskets, or broadswords, or canals, or statues, or songs." She believes that being able to do something with her career that brings her happiness equals success. She also cautions others not to define "success" too narrowly. Instead, one should continually assess one's strengths and weaknesses while seeking opportunities for feedback and ways to develop new skills. Remember that these opportunities may come from outside of your current position!

She considers her breadth of experience across Rohm and Haas her strongest attribute today. She can help with any type of problem because of her broad network of contacts and can intuit what issues might be significant for different groups. She knows where to go and whom to include either for troubleshooting directly or in terms of influencing. Cheryl also believes that she has solid credibility with many departments, because of her broad experience within the company.

Cheryl does not think that she needed to make any substantial sacrifices to achieve the success she has today. She believes that because life constantly changes, one needs to be flexible in order to constantly evaluate what is important and how to best attain what one wants. She states, "I guess you could call this compromise, but I think it is just reality."

Work–Life Balance

Finding a good work–life balance is a constant challenge for Cheryl. However, she has always tried to blend "life" into the challenges of work. For example, because travel is a passion for her, she tries to take the time to arrive at a business trip location early to visit the surrounding area. Cheryl has had the pleasure of visiting such places as Singapore, France, Japan, and Scotland by following this philosophy and she has even studied French.

Cheryl has also remained very active in the ACS and on boards or committees of non-profit organizations such as the Mural Arts Program, the Corporate Alliance for Drug Education, Women's Way, and the Philadelphia Distance Run. She notes that recognizing that we are part of a broad community always helps her keep perspective on what is important.

Diversity

Being a woman in fields still primarily consisting of men is also a challenge, according to Cheryl, and she is not just referring to the scientific field. Wallstreet has even less gender balance than chemistry does today. Working in the business world and investor relations means dealing with assumptions from people who have not often dealt with women in such roles. However, this is also a great opportunity, Cheryl believes. She says that by bringing a different perspective to the table, one can often take advantage of that "a-ha" factor (like watching the light bulb come on) for great success. This opportunity applies to anyone with a different perspective and supports the argument for increased success by increased diversity.

One point to remember about diversity is the importance of respecting cultural differences that arise in the business world. Some cultures have very different philosophies about women in business. Cheryl advised colleagues to do their homework when dealing with a different culture and to enjoy the challenge of performing the job well while still respecting cultural decorum.

Mentors

Having diverse mentors can benefit one's career as well. Cheryl has had many mentors: undergraduate and graduate advisors; various members of the Rohm and Haas community, both technical and business; and also those outside of the company or even the scientific community. She says it is useful to have a variety of advisors to provide a broad perspective and help you see the big picture. Good mentors will also show you how your skills fit into areas beyond your existing position.

Cheryl believes that good mentors are those with a broad knowledge of useful skills and an understanding of people. They should also have a solid knowledge of areas beyond their current work environment and the ability to look beyond the mentees' current position for opportunities. Cheryl advises that to find a good mentor, you should talk to everyone and be sure to look outside your own company. You should also go out of your way to listen to and meet those who have a different way of working and then ask them how their career has developed and what were keys to their success.

Leadership

Being able to integrate the business goals with the strengths of each employee is the sign of an excellent leader, according to Cheryl. She says that leaders should be able to successfully blend these two aspects into goals for each employee in a manner that is rewarding to them and meets the organization's needs. Leaders should also possess inquisitiveness and openness to what can be learned from other fields or disciplines. Additionally, as a successful chemist or engineer, one should have the ability to frame good questions, understand data in light of ambiguity, and look beyond the current problem to see the big picture and anticipate future issues.

Cheryl also agrees with most people in industry that effective communication skills are critical for everyone. However, communications is not something that is often taught. Vital to successful communications is understanding the interests and needs of the people with whom you must work. She personally always tries to empathize what it is like to be in the other person's shoes. She asks herself what might they be worrying about and what is important to them. This has helped Cheryl identify common ground and mutual areas of success for both parties. Additionally, Cheryl notes that women often have a strong understanding of how to present things to employees or have more empathetic communication styles.

Final Thoughts

Cheryl knew early on that she did not want to work in the laboratory forever, but she could not imagine the diverse path her successful career would take. It is always important to leave your options open when determining your next career move. Cheryl's final advice is, "...understand what you want, don't be afraid to acknowledge that your interests are changing, and accept new challenges!"

Chapter 15

Rebecca Seibert, Technology Manager: Making Choices

Jody A. Kocsis

Technology Manager, Engine Oils, Lubrizol Corporation, 29400 Lakeland Boulevard, Wickliffe, OH 44092

Success requires compromises. Or perhaps it is better stated that success requires one to make choices. Rebecca Seibert is a successful Technology Manager for Crompton Corporation. She believes that our career goals are a significant part of our life goals and if one keeps their life goals in focus, success will come in the right form and will require making choices. Rebecca has had to make choices, some easy and others not so easy.

Rebecca Seibert did not begin her education with the intent of becoming a chemist but rather a medical doctor. She was in an accelerated program at college and one semester away from medical school when she woke up one day and just knew that was not the direction she wanted to take with her life. She had a really "cool" chemistry professor at Gannon University named Gerry Andre and enjoyed his classes immensely. He sparked her interest in chemistry and physics, so she switched her major.

Rebecca earned a B.S. degree in Chemistry, from Gannon University in Erie, Pennsylvania, and an M.B.A. with an emphasis on Marketing Strategies from the University of New Haven in Connecticut. Rebecca also attended the University of Pittsburgh from 1986 through 1987 in the Ph.D. candidate program but left the program to work for Uniroyal Chemical Company.

Rebecca had made a life choice. Her fiancé was unsuccessful in getting a job in Pittsburgh and was hired by Sikorsky Aircraft in Connecticut. After a year of flying or driving eight hours every other weekend to see each other, they decided this was not the way they wanted to spend the next four to five years while Rebecca completed her Ph.D. Rebecca sent out resumés and within a month received two job offers.

Matthew (kneeling), Nathan (standing), Sean (in lap) Rebecca, and Emily Seibert (Courtesy of Rebecca Seibert.)

One of these offers was from Uniroyal Chemical Company (which later became Crompton and Knowles, then Crompton Corporation through a merger with Witco Corporation) as a Product Development Chemist. Rebecca performed this job for three years, became a Technical Sales Service (TSS) Research Chemist for four years, and then moved into Specialty Chemicals as a Project Manager. The last position involved commercialization of new products, as well as developing and implementing a new product and process development framework for the Specialty Chemicals business. After three years in this capacity, Rebecca made a career change to become a Commercial Development

Manager for the Petroleum Additives business. Her primary focus was on the petroleum area and involved design, implementation, and management of the business's portfolio as part of Crompton's technology portfolio management process. Rebecca was promoted in 2000 to her current position, Technology Manager, Petroleum Additives.

She has been fortunate in her career path. She has had opportunities to take on positions at Crompton that she has enjoyed and has been able to chart her own path. She never would have expected that she would hold the position she has today without a Ph.D. Fortunately, she has a supervisor who values her track record and knows how hard she works.

Rebecca's current role includes managing the Petroleum Additives technology portfolio, as well as managing a group of synthetic chemists ranging from technician to Ph.D. scientists. Rebecca's group focuses on new molecule and custom molecule synthesis for lubricant and fuel applications. She says that the technicians are just as creative and have patent histories as rich as many of the degreed chemists in the group. This reflects the atmosphere that Crompton has worked to achieve. It's not the degree; it's the talent that Crompton hires.

Her job is very diverse and ever-changing. Rebecca gets to keep her hand in the technical issues yet still handle program and people management issues as well. She also maintains the customer interaction that she really enjoyed when she was in technical service but doesn't have to travel as much. Finally, she enjoys supervising a group of really innovative, smart chemists.

Sacrifice or Opportunity?

Rebecca doesn't like to think in terms of sacrifices. She thinks we all make choices in life that can take a person in different directions. She has made many choices in her career, and, looking back, she thinks they have been the right choices for her at the time. Rebecca says, "Choosing and making the right decisions in one's career is as much about what your brain tells you is the sensible thing to do as well as what your heart tells you feels right."

Rebecca's family is the most important part of her life. They are her top priority, and they keep her balanced. She has had opportunities along the way to take on sales positions (presumably the path to business management and more financial reward) as well as offers to join consulting firms because of her experience in project management and portfolio management. Perhaps she will consider these options later in life. However, while she has four children to raise and nurture, she will seek positions that limit the traveling so she can spend more time with her family.

Balancing a Career and Home Life

"It's hard work!" Rebecca has been married to a wonderful man for 16 years and has four children. They are fantastic kids that she is extremely proud of and she really enjoys their company "as people." Her office at work is plastered with their photos and artwork. Rebecca believes children are our greatest assets for the future. She tries to keep her priorities straight, although she admits that some days she has to remind herself to "turn off" work when she gets home. She works all day, and when she arrives home she immediately tackles the challenges of children's homework, extracurricular activities, getting dinner ready, and preparing for the next work and school day. Personal, alone "down time" is very important and is probably the most difficult thing for her to fit into her schedule.

Rebecca really believes in giving time back to the community, and as a result, she and her family are very involved with community, school, and church activities. Her involvement with the church is the piece of her life she works hard to keep centered. All of these activities really help her keep that "balance" in her life. Of course, her husband and she must keep their calendars very up-to-date. She does not have the opportunity to cook elaborate meals very often, and her house does not get cleaned thoroughly every week! She states, "As the kids have gotten older, they have started to help out a lot with the household chores. This helps. I really do believe I perform better at work because I have a life outside of work that is important to me."

Her "therapy" is sewing and quilting. She loves to sew and create custom clothing and other items for people. Rebecca especially enjoys making baby christening gowns from wedding dresses. When she is stressed, she can be found at her sewing machine creating something.

A Mother and Professional

Emotionally, the hardest challenge Rebecca faces as a mother and a professional is the guilt that never really goes away over going to work rather than staying home while her children are young. Intellectually, she knows it is natural to feel guilty and she knows her children are in terrific shape. She rarely misses school events; she is there for the homework and extracurricular activities. However, there is a piece of her heart that is torn out each morning when she takes her youngest son to daycare. Certainly Rebecca would rather stay home and play with him because he is fun to be with, but she also enjoys the mental challenges of her job. Her children understand that her career is important, but not more important than them. Also her husband is very supportive of her career, which is essential and invaluable.

Logistically, Rebecca does have to give up weekends and evenings to travel for conferences and customer visits in her job. Time is always at a premium. She works with her children's teachers, coaches, and other people to make sure they are always up to date with events. Her husband and she have a rule that they never travel at the same time. That way, one parent is always at home during a workweek. She tries to plan her travel schedule around all these things. It doesn't always work, but most of the time she is home for the important school and extracurricular events. She keeps her supervisor informed of important dates as well, so he doesn't schedule a trip for her when she needs to be home.

Traits of a True Leader

Rebecca has a quote she uses as part of a team training module for new product development teams. This training module focuses on methods for consensus team decision making and the roles of program leaders. The quote is from a Chinese emperor who said, "A 'good leader' is one where the people say 'he did it'. A 'true leader' is one where the people say they did it themselves." In other words, leaders are people who can create the vision, communicate it clearly, and work hand in hand with the team to create, by consensus, the objective and the plan to achieve the goal. The best leaders lead by example and can be looked up to as the type of person one would like to become. A leader should be creative, energetic, ethical, self-motivated, resourceful, organized, a planner, and a problem solver. Also, a leader should be a good people manager in order to manage different personalities, as well as a good listener, and be able to quickly resolve issues whether they are people- or program-related. Good leaders need to be able to balance the "big picture perspective" as well as understand the details.

Removing obstacles for the people in her group so they can do their jobs is really important to Rebecca. She tries not to micromanage their activities unless given a reason. She believes a leader needs to learn what it takes to energize, motivate, and properly reward the individuals that work for them. Finally, true leaders needs to know how to have fun with their co-workers and need to recognize important events in their group's lives. "Remember they all have lives outside of work too," says Rebecca.

Chemistry Yesterday, Today, and Tomorrow

Chemistry, for the most part, is a male-dominated field. That fact has not changed since she started working 16 years ago. Rebecca has faced particular challenges as a woman in the field of chemistry. When she started in tire technology at the beginning of her career, both her youth and her gender could have been deficits. It took hard work, enthusiasm, integrity, and dedication to earn the respect of some of her co-workers, lab technicians, and hourly personnel that worked in the production plants. She was a 24-year-old female entering an atmosphere of all males who had 20 to 30 years work experience. In the end, she learned from them, and they learned from her. "They were a great group of guys who taught her a lot."

Rebecca spent a week at a leadership workshop a few years ago looking for consensus building tools for implementing Crompton's New Product and Process Development (NPPD) and Project Portfolio Management (PPM) processes. One of the most interesting and completely unexpected "learnings" she brought back from that week was a statement made by one of the coaches: "Ladies, even though you work in male-dominated industries, don't ever hide or give up your femininity. It is one of your strengths, one of your assets. Celebrate it." This really resonated for her at the time.

Definition of Success

Success is a very individualized concept. For Rebecca, success means doing the best she can in a career that brings her enjoyment and challenge. This is really important because, with retirement ages rising, a person's career can last more than 40 years. That is a long time to be in a career that one does not get excited about every morning and welcome the challenges of the day ahead. Success for her must include maintaining a balance between her career activities and other activities in her life. This doesn't mean she doesn't bring work home with her, because she does. However, she works to make sure it doesn't take over and steal from family time. The one piece of advice she would offer younger chemists is to keep long-term goals in mind—not just career goals—but life goals. If you keep those in focus, success will come in the right form. If you do your best even in the smallest of tasks, are persistent and inquisitive, always work hard and ethically, learn how be a team player, never stop learning new things, develop the habit of approaching problems positively and creatively, and strive to help others, you will be successful and happy in your career. To quote Helen Hayes, "Always aim for achievement, and forget about success." Success is not one single event, but thousands of little victories and accomplishments along the path of one's life. Does success require compromise? Yes, success requires compromise! Or perhaps it is better stated that success requires one to make choices.

Mentoring

Rebecca's mom is her mentor. She taught first grade for 30 years and was Rebecca's first role model as a working mother. She took a lot of pride in her job and really cared about the success of each and every student she taught.

Additionally, Rebecca's first supervisor at Uniroyal Chemical was a woman. She was a really tough supervisor that inspired Rebecca to blossom professionally. She has served as a mentor for Rebecca, who still goes to her to discuss ideas and options as they pertain to career and family issues.

Rebecca also tries to mentor younger chemists who come to work at Crompton. Just being available to listen to their concerns and help them find the channels to answers they are seeking is something really important that she tries to make time for. She believes mentors are people that want to make a positive difference in people's lives, to help other people do their jobs better, and to help others grow in their careers. Sharing and networking, both internally and externally, are great ways to find mentors and to informally mentor others.

Mentoring students is important too. For example, when children in Rebecca's town see her, they know her as Matthew's mom or Emily's Mom, who comes into the classroom with fun experiments for them to do. They see her as a "normal" person who works in a scientific field, not some "mad scientist" as they sometimes joke about with her. What they take away and carry home to their families is that science is fun and a viable career choice. Teachers see her and others at Crompton as resources for making their lesson plans more effective.

Career choice is very important. Rebecca says to pick a career that will challenge you, keep you engaged, and will be fun. Learn all you can. Learn the business, learn the chemistry, and learn whatever you need to do the best job you can in your current position with a mind toward future options. A person needs to network to meet and develop relationships with people throughout the whole organization. Find a need and fill it, and always do the best job you can. Don't be afraid to try something new. Research it, learn about it, and do it. One thing she has been told quite a bit in her career is "that can't be done" or "we tried that before and it didn't work." If someone tells you it can't be done, don't believe it, because you may be the one person with the talent and tenacity to *get it done!*

Choices

In conclusion, Rebecca Seibert provided these final thoughts on career choices. Look for a field that excites you and will challenge you for many years and is something at which you think you can excel. Don't let other people tell you not to go into the field because it doesn't pay enough, the hours are too long, it's too competitive,…whatever the reason. Money isn't everything; it just pays the mortgage and buys the groceries. Choose a field that will make you want to get out of bed in the morning and tackle that next challenge ahead. Choose a field that will let you attain your life goals, whatever they may be.

Chapter 16

Grace Torrijos, Owner and President: Business as a Family Affair

Rita Majerle[1] and Elizabeth A. Piocos[2]

[1]Hamline University, 1536 Hewitt Avenue, St. Paul, MN 55104
[2]Clairol Research and Development, Procter & Gamble Company,
2 Blachley Road, Stamford, CT 06922

This profile highlights a successful chemist who took her love for science from teaching to environmental testing to owning her own business. One way she has met the rigorous demands of such career moves is to integrate her entire family into her work. Grace Torrijos makes helping people and helping the environment a family affair.

As the owner and president of Solar Environmental Services, Inc. (SES), Grace Torrijos finds her life in a healthy balance, surrounded by the work she loves, her family, and a beautiful northwestern location. Anchorage, Alaska, is home to Grace and SES, an environmental testing, engineering, and consulting firm she founded in 1991. SES specializes in abatement solutions, project management, and testing for asbestos, lead, and other hazardous materials. Analysis for the materials of interest can occur in a variety of matrices, resulting in anything but routine analysis procedures. The projects can be really varied and may include determination of hazards from underground fuel tanks, establishing the presence of unhealthy environments in buildings, conducting surface and groundwater monitoring, and project management for abatement work.

"For example, often our job can be to examine the structure of buildings and test for the presence of any harmful substances. If substances are detected, we determine whether they pose a health hazard and require removal," explains Grace.

Grace Torrijos (Courtesy of Grace Torrijos.)

Education and Early Career

Her interest in chemistry was sparked in high school, leading Grace to the decision to pursue a degree in chemistry. Upon obtaining her bachelor's degree in 1979, she started teaching at both the high school and college level at Jimel Academy in the Philippines. Then Grace's family immigrated from the Philippines to Alaska in 1985, and she obtained a position as a lab instructor in analytical chemistry at the University of Alaska, Anchorage. Grace later joined a local environmental firm, Pittsburgh Testing Laboratories (PTL), and found herself learning microscopy to analyze and monitor asbestos samples. Though trained as a general analytical chemist, she became very proficient at microanalysis and rose to the position of director at PTL.

Although she was in a prominent position, Grace thought something was missing. It was during a fishing and camping trip with her husband and two young sons (who were ages 9 and 4 at the time) that she decided she wanted to start her own firm. "I was on this hunting and fishing trip, in a camper, meditating about my life and what I was doing, when I realized, I can do this for myself! I really enjoy my work." She told her family of her decision, came up with a name for the company, found an office when they returned from their weekend trip, and gave two weeks notice to her then current employer.

The Business

Today, Grace's staff consists of chemists, professional engineers, industrial hygienists, safety engineers, and certified inspectors. Laboratory services include phased contrast and electron microscopy analysis, and lead and asbestos assessment. Also included are analyses of various environmental pollutants, including PCBs and hydrocarbons, and ground and surface water monitoring, along with microbial analysis.

Solar Environmental Services, Inc., serves a wide variety of public and private clientele. Although SES is a relative newcomer to the environmental scene, the reputation of the firm is widespread and most of Grace's clientele come from referrals. Her reputation has proceeded her. When asked what is necessary to develop a clientele in a relatively competitive small business territory, Grace said that it is extremely important to keep current with regulations and techniques of analysis. "Continuing education and updating techniques and instrumentation are critical."

Work–Life Balance

From its inception, SES has truly been a family affair. Grace is from a business family. Her father, an engineer, helped her get a start in the business. Though recently retired, he played a key role in developing laboratory procedures and conducted laboratory analysis. Grace's husband does field work and is EPA-certified (Environmental Protection Agency) in lead analysis techniques. Her oldest son is currently studying computer programming at the University of Alaska, Anchorage. He is also trained in and has conducted mineral analysis and has developed a program for data correction. Her younger son is also working for the firm, performing data entry and results analysis.

Although Grace truly loves her work and her company, she advises keeping a close watch on the amount of time spent on work. "Don't permit work to take over. Make sure you have time for yourself!" She carefully guards her hours and has developed the business acumen necessary to bring her jobs to completion on time, without sacrificing time with her family.

In her spare time, Torrijos enjoys the Alaskan wilderness as an active participant in hunting, fishing, and camping. She also enjoys music and plays both the piano and the guitar. "Music runs in the family too," she stated. Both her sons play piano, and the oldest plays guitar and composes too.

Success

Grace defines success by the quality of the work she performs. "Satisfaction is felt when you know that you have helped someone who required assistance. It is also satisfying to develop and maintain a fine reputation in the field that you work in and remain in business for a long time. Compromise is not necessary as long as you know how to balance your time and your money."

Grace believes that long hours, hard work, and dedication have brought her to where she is today. Family support has been critical in sustaining SES. "I have a deep sense of accomplishment and fulfillment. I have learned, too, to manage time more efficiently. Time management is the most important factor in this business. It helps me prioritize my work schedule while not damaging my family obligations and relations."

Mentors and Final Advice

When asked about mentors in her life and how they helped, Grace said that her parents were her most influential mentors. Her father, in particular, gave her lots of pointers in business and shared the workload. Grace also credits her high school chemistry teacher as being a strong influence. "Her teaching method opened me up to the wonders of chemistry and science."

Finally, Grace's advice for those interested in owning their own business is to investigate the possibilities thoroughly. "You must love your work and be prepared to devote a lot of time to it, if you are planning to start your own business. There are many advantages and disadvantages and you must carefully weigh all of them." Being the successful woman she is, it is easy to see that Grace lives by her words and truly believes in her values.

Chapter 17

Diane M. Artman, Marketing Director: Transitions

Jody A. Kocsis

Technology Manager, Engine Oils, Lubrizol Corporation,
29400 Lakeland Boulevard, Wickliffe, OH 44092

Making things happen, taking risks, and willing to make
sacrifices are just a few words that describe Diane M. Artman.
She has put her chemical engineering and M.B.A. degrees to
good use in a wide variety of positions. Diane is currently
Marketing Director at GrafTech, headquartered in Cleveland,
Ohio.

Diane Artman earned a B.S. degree in Chemical Engineering from the
University of Cincinnati, where she graduated Magna cum Laude. Diane started
her career at Harshaw/Filtrol Partnership, as a process engineer, where she
developed process improvements for manufacturing. After a year, Diane joined
the Lubrizol Corporation as a process engineer for the Research and
Development pilot plant. One highlight of Diane's career as a process engineer
was the development of computer models based on the kinetics of an alkylation
process. This process increased capacity 50%, saved $250,000 in raw materials,
and eliminated a capital investment projected at $800,000. While working full-
time in the pilot plant, she also earned her M.B.A. from Baldwin-Wallace
College in Berea, Ohio.

After four years in the pilot plant, Diane was promoted to program manager
in automatic transmission fluids (ATFs). In this position, Diane managed a $6
million testing budget to obtain customer approvals for ATFs. To utilize the
skills from her M.B.A., Diane later became an ATF product manager. As such,
she developed, coordinated, and implemented the business plan, which contained

Diane M. Artman (Courtesy of Diane M. Artman.)

the key strategies for profit/market share goals. She also executed a successful product launch with pricing strategies and promotional information.

In 1995, Diane moved into the Heavy Duty Diesel Segment as a commercial manager for the Americas. She led the commercial development for a product upgrade that resulted in market share growth of approximately eight points. Serving as commercial manager for these two business areas, she significantly strengthened the business by creating competitive advantages, resulting in profitable business growth.

In 2000, Diane was promoted to Global Manager of Engine Oil Technology, responsible for managing a decentralized international research and product development group. The group achieved tremendous success in developing new products while operating within budget guidelines of $24 million. The 16 members of this culturally diverse team were located in Cleveland, Ohio; England; and Japan. Clearly, utilizing cross-functional teams was critical to control the time and cost of bringing solution-based technology to market. Diane's high degree of personal accountability and skillfulness at being both technically credible and commercially savvy were important assets in this role.

Change is Good

To apply her skills and expertise in a new field of technology, Diane decided to make a career change in 2003. It was a difficult decision, but a great development opportunity in a leadership role. She is currently the Director of Marketing at GrafTech International Ltd. GrafTech International Ltd. is a global company with more than 100 years of experience in the carbon and graphite industry. They manufacture and provide high-quality natural and synthetic graphite and carbon products and services for: steelmaking, ironmaking, fuel cells, ceramics industry, aluminum manufacture, and electronics.

Success

Diane defines success in two ways, "At life, you are successful when your children say you are a 'great mom' and look up to you. In business, I feel successful when I see people grow through their assignments, contribute to the organization's success, and reach their potential. I really enjoy taking the undeveloped talent and nurturing it." Diane likes to make things happen for the people she manages. She has created opportunities for everyone, including oversea assignments, project challenges, and recognition.

Karen Allen, program manager at Lubrizol had this to say about Diane, *"Diane has helped Lubrizol succeed by taking a personal interest in other outstanding achievers and helping them achieve their potential. She works behind the scenes to make things happen for these 'rising stars'. She has definitely helped me, and I consider her my first true mentor. We need more people like Diane in the industry—people who make it easier for others to be effective."*

Leadership

A good leader has to have a vision and an ability to implement that vision with the help of the team, but for Diane leadership is more. A leader must be an example to others and walk their talk. A leader has to be an example of integrity and honesty. Leaders have to be authentic. When Joe Gibbs, NASCAR owner for driver Bobby Labonte, won the 2000 NASCAR championship he said, "You win with people." Diane embraces Gibbs' comments and believes a leader wins through the development of people and recognition of their skills and accomplishments. Leadership does not mean getting people to do their jobs; it means getting people to do their best.

Family

Diane attributes her success to having her family behind her one hundred percent. Diane has been married to Dick for over 20 years and has two children, Alex and Chelsea. Dick said this about his wife:

"Juggling career, family, and personal time is not easy! Sometimes sacrifices have to be made, and it is usually Diane that has to make them but success seldom comes without sacrifice. To make this work we make it a team effort and, while we don't see each other as much as we might like, we make our time together special. While we are apart I know that she is working hard and meeting her responsibilities. I am very proud of what she has accomplished and always enjoy saying, 'That's my wife!'"

Balancing work and family can be challenging. Diane says she keeps the balance by maintaining her priorities, staying organized, and having excellent home support. It takes teamwork to overcome any challenge. Her husband assists with the housework and does all the cooking. Family time is important to Diane and her family enjoys beach vacations together. Diane also enjoys watching NASCAR racing, especially her favorite driver, Bobby Labonte. Diane has even indulged in racing classes, as she has the need for speed!

Science as a Career

Her advice to women thinking about science careers: "GET OUT OF

YOUR COMFORT ZONE." The horizon is wide open. Look for opportunities to grow, take on challenging assignments, develop new skills beyond the technical, and volunteer for opportunities for customer interaction or international travel. Finally Diane lives by these simple fundamentals: Do what you say and walk your talk. Make things happen for people. People need to see you for who you are.

Chapter 18

Marion Thurnauer, Senior Scientist: A View from a Government Laboratory

Ellen A. Keiter[1] and Elizabeth A. Piocos[2]

[1]Department of Chemistry, Eastern Illinois University, 600 Lincoln Avenue, Charleston, IL 61920
[2]Clairol Research and Development, Procter & Gamble Company, 2 Blachley Road, Stamford, CT 06922

This senior scientist and former division director at a major government laboratory has devoted considerable time and energy to increasing the role of women in science. In her reflections on successful career paths, she highlights the importance of a supportive network and good mentors.

Thanks to the influence of her family, Marion Thurnauer learned to appreciate the wonders of the universe at a very early age. Her father had a particularly big influence on fostering her interest in science, as did an aunt who was an astronomer. When her aunt showed her the rings of Saturn through a telescope, Marion says she was "hooked". While considering her major in college, Marion initially thought she wanted to work on biological problems, but her future husband, then a chemistry graduate student, convinced her that if she studied chemistry, she would gain expertise in physical techniques and approaches that could be applied to the study of biological systems. While she was at the University of Chicago pursuing a Ph.D., she became so interested in chemical problems that her earlier thoughts about biology faded.

Interestingly, however, when she was a postdoctoral associate at Argonne National Laboratory, she found herself engaged in essentially what she had much earlier set out to do, namely, applying a physical technique (magnetic resonance spectroscopy) to the study of a biological process (photosynthesis). Drawing from her own experience, Marion advises young people to avoid setting their career goals so tightly that there is no room to see the many exciting possibilities available.

Marion Thurnauer (Courtesy of Marion Thurnauer.)

The Path to Leadership

Following her postdoctoral appointment, Marion joined Argonne National Laboratory as an Assistant Chemist. She gradually rose to a Senior Chemist position and, in 1995, was appointed Director of the Chemistry Division. She believes her involvement in activities related to women in science directly contributed to her appointment to a leadership position. At the time she was hired at Argonne and for the next eight years, she was the only female Ph.D. staff member of Argonne's Chemistry Division. That experience prompted her to look for ways to support and encourage young women interested in science. This was in the mid-1980s, which she describes as an opportune time to promote science and technology careers for women and minorities because many people were asking who would be practicing science in the year 2000. One of her accomplishments was to organize a conference, Science Careers in Search of Women in 1987. Originally an event for college women, it is now held annually for high school women. After the second conference, laboratory management expressed a desire to hold an annual conference organized by female staff members. Marion and a core group of women saw this as an opportunity to convince laboratory management that outreach and career development are intimately linked. For example, young women could not be brought to Argonne and successfully encouraged to be scientists and engineers if they observed only a few women in relatively lower level positions. The group worked for a year with the laboratory director's office to establish the Women in Science (currently Women in Science and Technology; WIST) Program. The program, initiated in 1990, is designed to develop outreach activities and to foster recruitment, retention, and promotion of women. Two immediate outcomes of establishing the WIST Program were increased visibility with laboratory management and development of a supportive network of women. Marion believes both factors played a part in her eventual appointment to a directorship.

Marion identified a number of traits that contribute to being a good manager. They include being decisive but open to new ideas, having good organization and communication skills, being a good listener, and being fully engaged but at the same time maintaining a level of detachment. An ability to prioritize is also essential because some issues that are not highly important can take an inordinate amount of time and energy. Finally, a good manager must be aware when it is time to move on because effectiveness in a position, for various reasons, has diminished. For a good article on this point, Marion recommends "Choices Made, Lessons Learned" (*AWIS Magazine,* Summer, 2001) by Linda Mantel, who was at the time President of the Association for Women in Science (AWIS).

The same traits are essential for being an effective leader in a scientific field. In addition, Marion points out, one must be a visionary who can convey one's vision both "up and down". She also believes it is important to recognize that leadership has different "forms" and "durations". She cites an example from a Leadership Skills Workshop and Networking Luncheon held at the 2002 American Association for the Advancement of Science (AAAS) annual meeting, which was summarized in the Spring 2002 edition of *AWIS Magazine*. One of the speakers at the workshop, in referring to Nancy Hopkins' initiation of a study of women faculty at Massachusetts Institute of Technology (MIT), said "Hopkins was unaware that she had taken a leadership position. She simply did what needed to get done." Thus, it is not always necessary to be appointed to a leadership position to assume the role. However, demonstrating leadership may lead to such an appointment. This certainly appears to have been the case for Marion.

Mentoring

Marion has had and continues to have a number of mentors who have not only advised her but also have provided a great deal of support and encouragement. When she became division director, she realized she had to learn quickly how to function in a new situation and did not hesitate to seek advice from individuals who were in similar positions or at higher levels. However, she acknowledges that mentoring is not always an easy process. The key, she believes, is learning how to translate valuable advice to one's own frame of reference; only after this process (the "click", the "ah-hah") can one successfully apply it.

Marion recently participated in a mentoring panel as part of a Survival Skills for Women workshop series at Argonne in which she presented the following points of advice for mentors:

- DO: Listen, observe, remain open ... DON'T: Offer quick answers
- DO: Adjust approach to obtain results ... DON'T: Expect that one size fits all
- DO: Be realistic and constructive ... DON'T: Sugar-coat
- DO: Enjoy the process ... DON'T: Be self-serving
- DO: Remember times change—build on your experience within the context of the present ... DON'T: Hang on to out-of-date practices

Marion referred again to the last "DO" to specifically emphasize the importance of being attuned to modern ways of dealing with recurring issues.

For protégés, Marion offers the following suggestions taken from an article, "Don't Trust Anyone Under 30: Boomers Struggle with Their New Role as Mentors", in the June 5, 2003, edition of *The Wall Street Journal*.

- Temper your cynicism; give your mentor a chance to help
- Avoid mentors who seek the limelight or who want to build fan clubs or add to their bank of IOU's

- It's OK to want to advance; be honest about goals and your mentor will feel less threatened

The article also gives some additional advice for mentors:

- Be curious about protégés; listen more than talk
- Expect nothing in return beyond watching protégés grow and develop
- Don't resent their ambition; yes they'll someday replace you

Marion adds that good "people chemistry" between mentor and protégé is also critical.

Development

To keep abreast of issues that are of interest to her, Marion follows articles in periodicals, such as those quoted here, *The Wall Street Journal* and *AWIS Magazine*. In reading these articles, she finds she often has two reactions. One is, "Wasn't this written before?" and the second is, "Much of this does not apply to my particular situation." Although it is true that many issues do not change, Marion has learned that current articles can provide updated approaches to addressing them. As for the second reaction, she reiterated her belief in the importance of translating information into one's own frame of reference. Only when an article or training course can be related to one's own experience will it become useful. This idea was reinforced by comments she received after her original interview in this series appeared in the *Women Chemists Committee Newsletter*. Different women said they found that particular—but varied—parts of the interview provided useful information. She believes it was a matter of whether her comments and approaches resonated with their own issues and situations.

Challenges

As she progressed through her own career path, Marion learned firsthand the value of having a network of women and the importance of a "critical mass" of women in a workplace. She believes that, in general, if the work climate for women were to be plotted over time, it would show a sawtooth pattern with an upward slope. However, she acknowledges that many women chemists still face challenges and barriers, citing in particular the recent study, headed by Valerie Kuck, on the status of women in chemistry at the nation's colleges and universities. Among the conclusions Kuck drew from that study is that "women aren't viewed as equal partners in the scientific community along with men and that goes into the issue of respect." Marion believes that gaining respect is behind many of the issues women face. Nonetheless, she remains optimistic concerning the current generation of women entering the science workforce as they appear to have a more positive self-perception than those in the past because it is no longer as unusual for them to choose careers in science.

When asked about sacrifices she has had to make, Marion said she prefers to think of what she has gained from her choices, not what she missed. She points out that one makes choices all the time and, depending on one's perspective, a choice that precludes a different path may or may not be considered a sacrifice. She likens it to the question "Is the cup half empty or half full?"

Among the personal challenges Marion has faced is the issue of time. When she was Division Director, it was important for her to remain involved in her own research but she had to give up being intimately involved with every aspect, which was for her a compromise. Today's constant bombardment with information via e-mail, fax, and the print media and instant accessibility have added to the challenge. Marion also finds it hard to turn off her mind when she leaves the lab. To help her achieve balance between work and life, she finds it very helpful to work out with some form of exercise every day and she makes a point of sometimes being out of contact.

Success and Advice

Marion's definition of success is having a sense of satisfaction with one's life. It also involves achieving both the easily defined goals (such as completing a task) and the broader ones, while maintaining one's own internal standards. Finally, it includes accepting and learning from mistakes and moving on.

For those looking for a successful career, she advises choosing something one enjoys or, even better, is passionate about. Those interested in science have a good chance to decide if they really enjoy doing research during their graduate school years. Graduate school is a valuable experience even if it leads a person to decide that she/he would like to apply the experience and background to something other than research. This is particularly true now that graduate departments are considering additional training to prepare students for careers beyond research, such as administration. For further reading on this topic, Marion recommends an article by Susan M. Fitzpatrick, "Yes, Virginia, There is Life Beyond the Bench," which appeared in the Summer 2001 issue of *AWIS Magazine*. As a final note of advice she urges those coming up in their careers to "follow your goals and dreams, but always remain open to opportunities that may arise."

Chapter 19

Susan E. James, Vice-President of Worldwide Regulatory Affairs and Regulatory Compliance: Chemistry around the World

Jacqueline Erickson

GlaxoSmithKline, 1500 Littleton Road, Parsippany, NJ 07054

This profile highlights Sue James. In her career, Sue has taken on a variety of challenges, which have led her from being a development chemist in the United Kingdom to a successful position as the Worldwide Head of Regulatory Affairs and Regulatory Compliance, based in the United States.

Sue James is Vice President of Worldwide Regulatory Affairs and Regulatory Compliance at GlaxoSmithKline Consumer Healthcare. In this role, Sue is responsible for Regulatory Affairs, Regulatory Compliance, R&D Quality Assurance, and the Environmental Health and Safety Groups in both the United States and the United Kingdom. Sue started her career by earning a Joint Honors Degree in Chemistry and Biochemistry from a university in England. After earning her degree, Sue considered going on to earn a Ph.D. and entering academics, as she had several opportunities; including an offer of a research fellowship. However, she decided to gain some industrial experience instead. She applied for and obtained a job as a development chemist at Beecham Products (now GlaxoSmithKline Consumer Healthcare), in Surrey, England.

Early Career

As a development chemist, Sue was responsible for formulation studies, analytical method development, setting up clinical studies, and transferring the

Susan E. James (Courtesy of Susan E. James.)

product to the manufacturing facilities. Her initial responsibility was in the analgesics group. However, the company had a rotation program, where Sue could move into different therapeutic areas such as cough/cold, vitamins, antacids, and topical products. The end goal of developing a product was to submit a registration that would allow a product to be sold over the counter (OTC). This was Sue's first exposure to Regulatory Affairs, and as the development chemist she was responsible for generating the data and formatting it into documents so that the Regulatory Affairs Group could then submit the registration.

After a few years as a development chemist, Sue applied for a position in Regulatory Affairs but did not obtain that position. However, she continued to work with the group, especially in the area of European and international registration of products. A few years later, Sue applied for another position in

Regulatory Affairs and obtained this position. In this role, Sue worked on a variety of regulatory submissions in England and Europe, including the prescription to OTC switch of Tagamet®.

At this time, Sue's husband, also at GlaxoSmithKline, was given a transfer to the United States. Fortunately, Sue's manager, who had responsibility for worldwide regulatory affairs, also had an opening in the United States. Thus, Sue became U.S. Regulatory Affairs Manager, despite an initial lack of experience with U.S. regulations. Sue had moved up the ladder and taken on additional responsibilities over the years, due to her success in each position as well as through various reorganizations within the company. When asked what took her to the level she is at, Sue replied "being in the right place at the right time, as well as taking risks and opportunities when offered."

Work Climate Changes

One of the questions posed to Sue was a question about how the work climate has changed since she first started in the industry. Sue described work as oscillating with peaks and troughs and that the oscillations occur at a much higher frequency now. The expectations are much higher, and doing things right the first time is now essential. However, much more flexibility around work hours, more accommodation of work–life balance, and more women in senior positions are available, according to Sue.

Success

Sue believes that "success is achieving what I set out to achieve, as well as doing the right thing in the right way." Success is a balance between achieving the corporate goal and the manner in which it is achieved. Sue does not really believe that success requires compromise, but it does require teamwork at the corporate level and clear prioritization at a personal level.

Sue also indicated that to maximize opportunities, you might have to sacrifice something. In her case, she relocated from England to the United States. She noted it can be tough, especially when you are so far from family and friends but that "to make it workable, you have to deal with sacrifices." Sue also noted that less than 50% of the people offered international opportunities take the assignment but that these opportunities can be especially valuable in a career.

Sue believes that a key trait for success is integrity, no matter what the position. Sue also indicated that a successful scientist should have resilience or determination and be detail-oriented in order to work efficiently. A successful manager should have the ability to see the big picture and have a strong focus on goals, mixed with degrees of empathy, understanding and flexibility. Sue also commented that a successful leader should have good communication and motivational skills, as well as demonstrating believability and technical competence.

For career success, Sue believes that delivering results in a culturally appropriate manner and choosing the right team members are two of the most important skills. For her own professional development, Sue reads the *Harvard Business Review* and books on leadership, as well as books and magazines relating to regulatory affairs. In fact, Sue recommended the *Harvard Business Review*, and a book by Larry Senn, titled *Secret of a Winning Culture: Building High Performance Teams* as excellent publications for those interested in leadership. She also networks, primarily through trade associations, but thinks that meeting business objectives and leading her team are more important uses of her time.

Balance

On the subject of work–life balance, Sue indicates that she sets clear priorities, and on a case-by-case basis she determines what takes precedence. Sue and her husband schedule vacations twice a year, during which she won't check her voicemail or e-mail. She also schedules time to attend school activities for her two children so that she works around those activities. Sue has also found a way to schedule time for herself. Because she is very results-oriented, she enrolled in a martial arts school. This activity requires class attendance to progress to the next level, and the exercise of punching, kicking, and yelling is good for relieving stress. Finally, as a manager, Sue also encourages others to set priorities and balance their work–life activities, as she believes it is important to lead by example.

Mentors

For the most part, Sue's mentors have been her direct managers. They have helped by providing clear direction and constructive criticism. "You should never underestimate the value of constructive criticism," said Sue. Her family members have also been key mentors. Her husband is very supportive. He has always helped her to keep a clear perspective and a balance at work and home.

According to Sue, some of the key traits in a good mentor are that they should be approachable and a good listener. They should also have good insights derived from life experiences and a high emotional intelligence. At the same time, a mentee should be committed, goal-oriented, flexible, and self-motivated.

They should have potential in their field and understand their responsibility for their own development, says Sue.

In order to find mentors, Sue advised students to do research and be inventive in reaching out and contacting people. She also suggested summer internships as a way to gain hands-on experience and find people interested in mentoring. Often the human resources department of a local company can find someone within the company to mentor a student, and some companies have high school outreach programs.

Final Thoughts

Sue believes in order to have a satisfying career, you should choose something you enjoy, for "nothing is more miserable than spending eight hours a day, five days a week on something you hate." In addition, a person should be open to trying new things. As your experience widens, you may find something that you enjoy more and are better at than what you initially started out doing.

Finally, Sue advises that when starting a career, you should have "positivity, persistence, and patience." By that, she means that you should have a positive attitude and approach, and deliver your best all the time. She also said "You need to be resilient and cautiously ambitious—and don't expect too much too soon and don't bite off more than you can chew."

Chapter 20

Rita A. Bleser, Vice-President for Research and Development: A Unique Path to Management

Ellen A. Keiter

Department of Chemistry, Eastern Illinois University, 600 Lincoln Avenue, Charleston, IL 61920

This article portrays an individual with a bachelor's degree in chemistry whose management skills have elevated her to a vice presidential position in a major chemical company. Rita Bleser is currently Vice President for Research and Development for the Pharmaceutical Division of Mallinckrodt. In addition to her chemistry degree, she holds M.B.A. and J.D. degrees as well.

It was the intellectual appeal of the subject that motivated Rita Bleser to choose chemistry as her major when she started college in the 1970s. Initially intending to enroll in medical school upon graduation, she recognized about midway through her undergraduate studies that a career in medicine wasn't for her. However, she also could not see herself spending decades as a bench chemist. In order to maximize her options, Rita elected to complete the American Chemical Society (ACS)-approved chemistry major as well as a combined chemistry and management track offered by her undergraduate school, Eastern Illinois University.

Those early choices set the stage for Rita's entry into the chemical industry and her very successful march up the management ladder. Along the way, the process of decision-making she used while still an undergraduate has been repeated often. And the same key elements have always been apparent: a keen awareness of her own interests and strengths, plus an openness to every available opportunity.

Rita A. Bleser (Courtesy of Rita A. Bleser.)

From Manufacturing to Law

Rita's first position as a chemist was with Lever Brothers where she worked as a floor supervisor in manufacturing and packaging soap. While there, she continued her education part-time, earning an M.B.A. from St. Louis University in 1980. She also developed an interest in patent law and saw it as a new opportunity and another decision point. In 1981, she resigned from her position with Lever and enrolled in Washington University Law School as a full-time student.

After completion of a law degree in 1984, Rita worked for two years in a St. Louis law firm specializing on litigation involving corporations. In 1986, she joined the regulatory and litigation division of Mallinckrodt, Incorporated. Over the next nine years she held various positions in the company's legal department and was eventually appointed Assistant General Counsel.

A Return to Manufacturing

In 1995, after being urged repeatedly by the company president to return to manufacturing, Rita accepted appointment as General Manager of Operations for Mallinckrodt's St. Louis Plant. She credits the president as having been a significant mentor over much of her career. He has recognized her strengths, helped her see new opportunities, and provided valuable support. Rita believes that the best mentors are at least two levels above those they are helping because the distance in position allows them to be objective and yet see an individual's potential.

From the remainder of her resumé, one can conclude that Rita's choice to make the switch back to the manufacturing area was the right one for her. In 1997 she was named Vice President and General Manager of Operations and in 2001 she became Vice President for Research and Development of the Pharmaceuticals Division at Mallinckrodt. Included in her current responsibilities are supervising synthetic chemists who are looking for new pathways to produce drugs, particularly generics, and overseeing a pilot plant. At the time of her interview, she had just returned from a trip to China to assess progress on a project she initiated.

Development

Although her titles and responsibilities have been quite varied, Rita describes her career as always involving science, which is one of the reasons she has thoroughly enjoyed it. She also values all of her experiences, both in education and employment. While the combined emphasis on chemistry and management in her undergraduate program certainly provided a foundation for her career, so did her first position as a foreman at Lever Brothers. The leadership and communication skills she developed there have proven valuable ever since; even in the practice of law.

126

And because of the heavy regulation required in the pharmaceutical industry, her background in law continues to be very relevant.

Throughout her career, professional development short courses and seminars have also helped Rita remain current and have prepared her for advancement. When she was being groomed for a vice presidential appointment, for example, she participated in a two-week course on management and leadership. Reading books on management is an important and continuous aspect of professional development for her as well. Although Rita is happy in her current position, she keeps her eye on the job market and maintains an up-to-date resumé. She thinks it helps her take stock of where she is professionally.

Balance

Rita does not believe her achievements have required significant sacrifices. She has a family that includes husband Steve (also a chemist) and their three children. When their first two were born, they both continued their careers and hired a nanny for help with childcare. With the advent of their third child, Steve elected to stay home for a year and, five years later, that's still the arrangement. While acknowledging that this strategy is not for everyone and may not be permanent for them, Rita describes it as a good choice for their family. She very much enjoys spending time with her children and counts "playing with them" as her principal leisure-time activity. Favorites include shopping with her 14-year-old daughter, playing catch with her 10-year-old son, and swimming with her 5-year-old daughter. Another of her favorite pastimes—sewing—is also centered on her children. On her recent trip to China, for instance, she bought some red silk fabric, which is destined to become a pair of boxer shorts for her son.

Rita also doesn't believe compromise is essential to success. In fact, she believes that if one is in a position involving a lot of compromise, it most likely means the person should look for something else to do. In her view, it's absolutely essential to remain true to one's own value system, one's "internal compass."

Reflections and Advice

Among the changes Rita has observed during more than 25 years in the chemical industry are the increased impact of the global economy, a stronger focus on productivity, a decrease in the number of ancillary positions, and a more business-driven climate. She has also noted an increase in the number of women in her company, especially among chemical engineers.

The career advice Rita offers to others begins with "Be true to yourself." In part that means choosing something that will allow you to be satisfied with what you do and how you do it. One of her key words is "focus". Remain focused on what you want to do and on whatever project or piece of work you're involved with at a particular time. Finally, striving to make a significant difference should underscore everything.

Chapter 21

Beverly Dollar, Chief Intellectual Property Counsel: The Joy in Chemistry

Frankie Wood-Black

ConocoPhillips, 6855 Lake Road, Ponca City, OK 74604

This profile highlights a successful woman who took her interest in science into the field of law. With a B.S. degree in Chemical Engineering and a juris doctorate, Beverly Dollar has seen many facets of intellectual property law applied to the chemical industry.

The official corporate biography of Beverly Dollar is very short and to the point. It somewhat matches her personality only in the sense that it is direct and does not mince any words. However, it cannot accurately portray her dynamic personality. The facts are simple—she was hired by Phillips Petroleum (now ConocoPhillips) as a patent attorney, just prior to the completion of her juris doctorate from the University of Tulsa, in 1988. Beverly had previously received a Bachelor of Science degree in Chemical Engineering from the University of Oklahoma in 1984. The only other highlights mentioned in her corporate biography are that she was associate general patent counsel beginning in 1995 and served as a staff attorney until the ConocoPhillips merger. In that capacity she provided support in the finance and human resources areas in the general counsel office. Once the merger was completed, she became the Chief Intellectual Property Counsel.

This simple unassuming biography does not provide you with the insight that she is fun, driven, and enjoys her work and life. She has a disarming personality in that if you are negotiating a deal or a settlement with her you had better be on your toes, as she is a competitive, fair, and ruthless negotiator. Beverly is a true success story and balances her job, family, and lifestyle to do what she enjoys while still making it to a very senior position fairly early in her career.

Beverly Dollar (Courtesy of Beverly Dollar.)

Getting Started

The path to the Chief Intellectual Property Counsel really started when Beverly went to college. She has a wide number of interests, including a fascination with science and a love for learning, through which she is always seeking a challenge. Although she enjoyed science, Beverly initially thought that she wanted to be an architect. When architecture wasn't scientific enough, she began searching for something better-suited. She ultimately opted for Chemical Engineering. The reason she gave is that they got to brew beer as a class project, and it seemed closer to having real-life applications compared with other science disciplines.

When asked what took her to where she is today, Beverly tells a story similar to the one in which she ended up in Chemical Engineering. She has an interest in everything and a willingness to give new things a try. She has done a number of outside projects where she takes on the challenge or the problem and does not necessarily worry about whether or not it is in her current job description. This philosophy applies outside the office for her as well. Her key criteria is that the problem or project be interesting and potentially a bit outside of the box. Beverly describes herself as impatient, with a short attention span, and kind of a "grass-is-always-greener" type of person. She also says that she has a tendency to take on others' problems, which makes sense for someone who ultimately became a lawyer.

As for the traditional rungs of the career ladder, Beverly started as part of the corporate legal staff in the Intellectual Property Department, which during her career became the Patent and Licensing Department, and then ultimately changed back into the Intellectual Property Department. She did branch out of the intellectual property field for a short time as a staff attorney working on finance and human resource issues. Over the years she has seen the intellectual property, patent, and licensing departments swell and shrink depending on the corporation's strategic view and industry trends at the time. Today, she manages the Intellectual Property Group in corporate legal, which consists of 31 people.

Mentors

Beverly indicated that she has had a number of mentors along the way, from her initial supervisor, to the past Chief Intellectual Property Counsel. She did indicate that there were not many female figures that she could approach at the time. However, she still worked with her mentors to develop the necessary career skills she needed to achieve the position of influence that she has today.

Work–Life Balance

Beverly indicated that she has had to make a number of trade-offs. For example, she spends less time at home because of travel requirements. In addition, she also indicated that she doesn't necessarily view her home life as traditional, although she is married and has two very active children.

However, Beverly has refused to sacrifice her sense of humor. She says that she really tries to put people at ease and others say she does a wonderful job of it. She uses this skill to her professional advantage. From what others have seen, Beverly does not really let anything stop her from enjoying her work and her home life. She has a sense of spirit that inspires the attitude: "If you want to let the workplace get you down it will. If you make it fun and don't take things too seriously, you can balance and prioritize things." Beverly's sense of fun does not get in the way her work; she just has found a way of making work more enjoyable.

Beverly's sense of humor is also critical to her balancing act. She indicated that humor is a requirement for dealing with the second important part of the balancing act—knowing when to sacrifice perfection. It always comes down to how you make some of those choices. For example, when you can't be the perfect June Cleaver and make the treats for the third grade class but can provide them by bringing the decorated cupcakes from the local grocery store. A support group (spouse or family) is a must in this situation as well. Beverly also believes that balance is not necessarily like leveling the scale but at times is more like a pendulum, where you consciously pick the times when you focus on work and when you focus on family. She says that you must pay close attention to know when those pendulum swings need to occur.

Changes in Work Climate

Working in a major corporation, which has undergone a lot of change since she started in 1987, Beverly has seen a number of changes first hand. The work environment when she started at Phillips Petroleum was heavily influenced by the takeover and merger mania that was occurring in the mid- to late-1980's. The company at that time had to begin making changes that were more pragmatic and survival-driven. Beverly also points out that today's work environment seems to be much more reactionary than when she first started. In the past, she indicated, people seemed to have more time to focus on the task at hand and think about how they would pursue a specific problem or initiative. Today, everyone seems to be more driven by the here and now; what is important at this instant. The merger with Conoco brought a term into the corporate vernacular, "at this 10 seconds", which seems to highlight this shift in the perception of time.

Another major difference in the workplace is that of change itself. Change and the issues associated with change were present in the workplace, when Beverly first started with the company. Yet, today the pace and implications of change are much faster and more significant. This is exemplified by the adjustments that Phillips Petroleum went through from 2000 to 2002. Phillips Petroleum acquired ARCO Alaska in 2000, bought TOSCO in 2001, and in 2002 merged with Conoco to become ConocoPhillips. Those are significant changes, and the pace of those changes has left the workforce with more of a realistic view of life. This culture of pragmatism is not unique to the ConocoPhillips workforce but can be seen in a number of different companies that have gone through major changes, throughout the industry as well.

Success

How does Beverly define success? She has a relatively simple answer—getting paid for something that she enjoys. Beverly sees herself defined as both a mother and a professional and says that she could not have done it without a great family support group, led by a very supportive spouse. But, success to her is definitely in the enjoyment. Beverly says that she also likes being recognized for doing a good job, but that such a definition is too narrow. In fact, she believes that most people define success too narrowly and as such don't really understand their own intentions. A person really has to look at one's self and one's self-definition and to understand what one wants to achieve. Hence, it is important that she defines herself as a mother and a professional who enjoys what she is doing—and gets paid for at least part of it.

If you get to know Beverly, you will understand her definition of success doesn't really allow for compromise. If Beverly is not enjoying what she is doing, she is probably going to do one of two things: either find a way to make it enjoyable or do something else. Beverly has drawn some clear personal lines. She won't sacrifice her sense of humor, and she is self-confident enough to know when she won't do something to compromise her principles or values. Thus, for her, compromise really isn't an issue. That is why Beverly's final advice is to do something that you love because you are more likely to be good at it and to be happy while you're doing it.

Chapter 22

Sharon L. Haynie, Research Scientist: Laboratory Investigation and Outreach

Arlene A. Garrison

Office of Research, University of Tennessee, 409 Andy Holt Tower, Knoxville, TN 37996

Many chemists devote significant time and energy to mentoring students. This profile of a career laboratory chemist illustrates the impact of reaching out to disadvantaged high-school students. In the laboratory, the interviewee has developed products that integrate biology with chemistry to substantially improve the environment.

Sharon Haynie has demonstrated her lifelong commitment to mentoring and encouraging young scientists during her work at DuPont and her extensive activity in the American Chemical Society. Sharon grew up in Baltimore, Maryland, and received her biochemistry degree from the University of Pennsylvania in 1976. From there, she proceeded to the Massachusetts Institute of Technology where, in 1982, she completed her doctorate in chemistry. After spending a few years with the AT&T Bell Laboratories, she joined DuPont Chemical, where she has remained for 20 years. She has worked in a variety of research groups that reflect her interest in bioscience and chemistry, including the Vascular Graft Program and the Fibers Department Biomaterials Group. For the past 13 years she has worked in Wilmington, Delaware, relatively close to her hometown of Baltimore, for DuPont Central Research in the Biochemical Science and Engineering Department.

Tradition

Sharon's career is viewed by many as a traditional one. Her education was continuous, and she chose graduate study at a premier university. After finishing, she could have selected an academic career. However, research into

Sharon L. Haynie (Courtesy of Sharon L. Haynie.)

industrial materials captured her interest and she continues to focus on the basic science necessary in the development of commercial materials. Her love of laboratory work is recognized in DuPont, and she believes that her skills are best used in the creative and innovative research she performs at the bench.

Environmentally Responsible

Sharon's research at DuPont has extended to a variety of projects. Her recent and current projects have emphasized environmentally friendly processes that reduce environmental impact compared with earlier processes that can then be discarded. One recent project involves the use of biological catalysts to develop alternative products that are manufactured from biological feedstocks rather than petrochemicals. These new chemical feedstocks can be replaced in a single growing season. In many cases, safer processes can be developed that result in reduced waste.

To achieve the many goals of the projects, much of Sharon's work has involved collaborations and multidisciplinary investigation. Sharon notes that the quest to make better products to serve society requires extensive collaboration, not the single scientist in the lab that is often pictured. A recent example of the types of team projects she is involved in is an effort to convert corn sugar to an aromatic monomer. Throughout her career, Sharon has emphasized laboratory work and she loves directing bench work. Sharon views the integration of materials and biology as a challenge and an important research path for the future. She is comfortable with the challenge and believes that the end result will be new ways to create better products that benefit society.

American Chemical Society

DuPont has actively supported Sharon's involvement with the American Chemical Society (ACS), and Sharon has demonstrated extensive leadership in her volunteer activities. Sharon is very involved in the Philadelphia Section of the ACS and served as Chair of the section in 2003. She was elected councilor representing her local section in the national ACS organization for several years and was recently elected again to the 2004–2006 term. Council members serve as the primary decision-making body of the American Chemical Society. Sharon has also served on numerous national ACS committees. In addition to her personal interest, Sharon attributes her activity in ACS to a number of men and women role models who have provided great encouragement and support of her ACS activities.

Mentor

The ACS Project SEED (Summer Employment for Economically Disadvantaged) program has been of special interest to Sharon. The SEED program provides financial support and information about potential science careers to students. The program is designed to provide students with exposure to scientists and to the thinking processes of science. The students spend time in Sharon's laboratory, where they are exposed to much more than just science. To many of the students, the professional work climate and the intellectual stimulation have a much larger impact than learning about the specific science involved. Many of the students would be unable to participate in the summer experience if not for the financial support they receive while participating. Although some SEED participants pursue science careers, some choose business or other professions. The summer program provides a chance for students to learn the thought processes that are common to research, such as probing, critical thinking, and analysis. Sharon points out that these skills are important to all technically competent citizens, whatever career they pursue.

Sharon has served as a mentor to students in the SEED program for nearly a decade and often stays in touch with the students for many years. She believes that volunteer activity is an important professional responsibility. In particular, she believes it is important to be an accessible example to students and women who are in the early career stages. While Sharon emphasizes her commitment to helping others, she resists recognition of her extensive volunteer work, insisting that the opportunity to provide support to students is its own reward.

Work and Life

Work–life balance has not been a difficult issue for Sharon. She enjoys traveling and is often accompanied by her mother. Sharon does not believe that choosing to be a scientist has required any personal sacrifice, and she would definitely make the same choice again.

Sharon's immediate and extended family made the many sacrifices in order for her to achieve many privileges (e.g., quality education; choice in her profession). She thinks that the best ways that she can acknowledge and honor those sacrifices are to be mindful, to enjoy the privileges of performing one's chosen work, and to continually give back to others. Her contributions to science have been significant and personally gratifying. Sharon does not believe that she has been forced to make difficult choices in order to have a traditional chemistry career, and she continues to enjoy her work.

Chapter 23

Sharon L. Fox, Strategic Planning Consultant: Working in a Truly Global Capacity

Amber S. Hinkle

Bayer Material Science, 8500 West Bay Road, MS–18, Baytown, TX 77520

Sharon Fox has had a truly interesting career journey, from manufacturing chemist, through finance, then quality, then customer service, and on to strategic planning. However, her most interesting career move has been to accept an assignment based in Germany, where she is still currently located.

Sharon L. Fox earned her bachelor's degree in chemistry at Massachusetts Institute of Technology (MIT) in 1990. She then went on to study physical chemistry at the University of Texas in Austin, where she earned a Ph.D. in 1994. Bayer Corporation hired her directly out of graduate school, to work at a polycarbonate manufacturing site in Baytown, Texas. After spending four years in the production laboratory, Sharon moved to the company's U.S. headquarters in Pittsburgh, Pennsylvania. While there, she completed her Executive M.B.A. from the Katz School of Business. Sharon served in several roles over the ensuing years, in the areas of finance, quality, and customer service. In 2003, she accepted an assignment in Germany in strategic planning. When several Bayer business units combined and spun off, Sharon decided to join the newly formed company Lanxess and remain working in Germany for the time being.

Education and Early Career

When Sharon first began studying at MIT, she was not sure exactly which career path she would pursue. However, chemistry was the hardest class for her as she started her undergraduate curriculum. Therefore, she found that because

---

Sharon L. Fox (Courtesy of Sharon L. Fox.)

she was investing so much time into understanding how things worked, she could bring that knowledge back and help out her fellow classmates as well. The feeling of success this gave her, on both a personal and professional level, resulted in a sense of fulfillment that she found in no other subject. Hence, she continued on in the direction of chemistry throughout her studies and into graduate school.

Once finished with graduate school, Sharon found a very different environment in the "real world" of the chemical industry. She found that working at a chemical plant entailed such novelties as riding a bicycle through the plant to your office and wearing a hard hat and steel-toed shoes every day. Her new experiences also included working a night shift in order to learn the manufacturing process, since the plant never sleeps.

Working in the production laboratory, Sharon spent her time developing lab methods for quality assurance and for trouble-shooting the manufacturing process. She also implemented new instrumentation and worked on process improvement projects. As Sharon learned more about product development and had experiences such as coordinating a project to build a new lab at the plant, she became more and more interested in the business side of the industry. Thus, she began taking classes as part of an executive M.B.A. program, which she completed in 2000.

In 1998, Sharon had the opportunity to take a developmental position within the company as a financial analyst for capital projects in the Bayer Plastics Division. She made the move not only from manufacturing to finance but also from Baytown, Texas, to Pittsburgh, Pennsylvania. After this one-year developmental stint, Sharon was promoted to Manager of Quality Improvement and Certifications for the PolyChem business for two years. Sharon then went on to act as Manager, Polyurethanes Customer Service (domestic and export), for two years. These assignments gave Sharon an opportunity to use both her chemistry and business know-how and gave her a much broader perspective on the corporation as a whole.

Throughout her career with Bayer, Sharon had been studying German as a second language in order to better communicate with her foreign colleagues. Finally, in 2003, she was given the opportunity to really use her language skills when she accepted a position in Germany as a Strategic Planning Consultant in Corporate Development.

In 2004, Lanxess was formed and Sharon chose to go with the new challenges that were afforded in being part of a brand new company. She still enjoys being involved in strategic planning and also enjoys living and vacationing in Europe. Sharon says that her adventures with Bayer have been a result of several factors: hard work, good mentors (both formal and informal), and a willingness to take on positions that were outside her formal experience bases.

Foreign Assignment

As far as Sharon is concerned, taking an assignment in Germany for several years is a decision she will never regret. She highly recommends accepting such an opportunity should it present itself. Although living farther away from family and friends can be a challenge, she knows her time in Germany is limited and that she may never have such a chance again.

The knowledge and experience you can gain from working in a different country can be invaluable, according to Sharon. She has seen how her company operates in a different culture and thus how company decisions are made on the global scale. She has also greatly expanded her own network and can now more effectively interact with many of her peers from around the world. Sharon points out that this is especially important for Americans, who tend not to experience the wider world view as often. She also believes that participating in such an assignment shows your flexibility and ability to solve problems on a variety of levels, thus potentially leading to future promotions and certainly making you more marketable.

To this point, Sharon would say that personal time is the one area where she has made significant sacrifices. Long hours have been the norm, although even now she says she works on balancing her time better. However, she is also quick to point out the specific reasons for some of these sacrifices, which have paid off in wonderful ways. Despite working full-time, she completed her executive M.B.A. and took language lessons for several years, which both consumed much of her personal time. However, this commitment resulted in the fulfillment of one of her key goals: a chance at this international delegate assignment. Without the language skills and an M.B.A., Sharon believes she would not have been considered for a job in Germany. Now she has the unique opportunity to travel throughout Europe and interact daily in another language and culture. "Definitely a fair gain for the earlier sacrifice," she notes.

Success

Sharon believes that success in its most basic form is achieving goals that you set for yourself in life, be they professional or personal, which bring you satisfaction and fulfillment. "The question is really, what kind of achievements will bring you satisfaction and fulfillment? Do you want to be the technical expert everyone relies on? Do you want to be rich? Do you want people to follow you? Do you want to win the Nobel Prize?" asks Sharon.

For Sharon, success is defined by surviving each day and having contributed to the better good. She knows this sounds ultruistic but she firmly believes in leaving something better than before. When she can achieve project goals and add a better understanding or value for all, Sharon knows she is successful.

Although no one has the perfect equation for success, Sharon believes that the number one skill for advancement is convincing others that you can do it, even when you aren't sure yourself. She also recommends that you work hard to understand as much of the full perspective as possible as you work through projects; understand what is being asked of you and be sure to deliver just that, if not more. Keep in mind who has something to lose, to gain, where something could go wrong, what can you do to prevent it from going awry, and what will be your Plan B if it does. Have your facts together, and defend your recommendations professionally, says Sharon and you will go far.

Sharon also notes that success always involves compromises but not necessarily ones that are unacceptable. "The benefit of arriving at the goal must outweigh any sacrifices made achieving the goal. Otherwise, despite having achieved the goal, you weren't successful." However, Sharon would not regard some areas for compromise, such as legal, ethical, and moral obligations. Clearly, not all people have the same view, and at times such beliefs are tested. However, the consequences of such compromises can be devastating on both a personal and professional level, as we have all surely witnessed in recent years. When it comes to family and personal time, one must decide what is more important and whether any compromises could be made. By being creative and flexible, you can sometimes have the best of both worlds, according to Sharon.

Again Sharon mentioned that her personal time is valuable to her but that she was willing to give some of it up for the benefits of her job. She thought her company valued long hours. In fact, the eight-hour day no longer exists as often. Fortunately, for Sharon, the longer days were rewarded successfully in that she could contribute as much as she desired.

Mentors

Mentors have been a great help to Sharon, both from a guidance perspective as well as for making sure her name was visible and considered when opportunities were discussed that she was unaware of. For Sharon, the most beneficial aspect of being mentored was feedback regarding the impression she leaves people with and recommendations on how to strengthen what she does well. Her mentors also advised her to work on approaching some things differently and how to make her points clear without leaving any negative impressions. Another positive impact of Sharon's mentors was assistance in understanding better how the organization functioned and how to approach some tricky personnel situations.

Good mentors have provided Sharon with valuable information based on their organizational experience as well. She also recommends that a mentor be in a position at least one level above yours but still approachable. Additionally, good mentors are concerned about the development of the people they guide and coach. In looking for a mentor, you may want to ask yourself whether colleagues have been promoted from under this individual. Finally, Sharon recommends avoiding any potential mentor who lacks discretion when discussing other colleagues when they are not present; a sense of trust is crucial between mentor and mentee.

If a formal mentoring program is unavailable, Sharon suggests identifying someone whom you believe has been successful in a direction that you think you would like to follow (scientifically or managerially). Look for the traits Sharon described. She notes that an effective mentor need not be someone in your department, (in fact, it may be better if not), but that person should be relatively accessible to meet. Approach the individual, explaining your intention to seek guidance regarding upcoming decisions or professional situations. Ask whether that individual would be willing to meet and share perspectives or approaches that have been successful. You will probably know right away whether this person would discuss issues or not. If the discussion goes well, chances are you can periodically return to the individual and gradually develop a mentoring relationship. Sharon reminds us that not all successful people are interested in being mentors, and not all make good mentors, so do not be discouraged if the first few approaches do not work out.

Professional Development

Sharon estimates that she spends about 2% of her time on her own professional development. Recently, that has entailed taking courses in strategy for business development and reading various books. For studying the behaviors of groups and leadership, Sharon recommends Machiavelli's *The Prince.* Regarding moral and ethical beliefs, as well as general confidence building, she recommends Wu Wie's *I Ching Life: Living It*, which Sharon admits might sound weird to some but she really enjoyed it. She says that before she read *I Ching*, she avoided networking with people she did not respect. However, this book shared insight on how we all need to work closely with a variety of people, even those we do not respect. Of course networking with people does not mean you have to socialize with them outside of work either. This book also reminded Sharon that she should not give herself less credit than she deserves, which she thinks she and many other women have a tendency to do. She says we should learn from our mistakes but realize they are not all really mistakes. "Remind yourself that, while something could have been done better, you probably did it better than some others, too. Mole hills really are mole hills."

Leadership

Good listening skills and a sincere interest to develop and promote the people working for you are key to being an effective leader, according to Sharon. She believes that these skills are usually relatively easy for women to develop. The skills that are equally important, but not typically as easy to pick up, center around matters such as effective conflict and expectation management and understanding when to use collaborative problem solving (let's work through this together) versus directive problem solving (I need this in 1 hour and done a specific way). Additionally good leaders should look for these skills when hiring people: good technical methodology and cross-checks, good listening skills, and a willingness to challenge the accepted ways of doing something but in a professional manner. Sharon reminds us that how the challenge is delivered is as important as the challenge itself in order for the message to be received. She also recommends that leaders and their team members take advantage of the many feedback tools that exist for soliciting feedback from peers, supervisors, and direct reports.

Work Climate

The work climate has intensified, with far more focus on cost-cutting, over the years Sharon has worked for Bayer and now Lanxess. She states that for the first time since she has been with the company, significant lay-offs have occurred and even more are planned. Also the spin-off of Lanxess is something that very few believed would truly occur, based on Bayer's beginnings in colorants and chemicals. Sharon predicts that the future should be quite interesting.

Most of the challenges she has faced personally have not been gender-specific. Rather they were challenges anyone working to establish themselves would face. Sharon did note, however, that women do need to be more careful than men regarding dress, personal behavior, and general work ethics as there is certainly still a double standard in existence, but she does not really consider that a particular challenge.

While not directly impacted herself, Sharon does believe that discrimination still happens against women but it is hard to recognize sometimes. For example, if you don't have a lot of people working for you, you may not see the pay differences. Also women are partially discriminated against when considered for positions in male-dominated areas, because they have different personalities and react differently to situations than men do. There is still some worry as to how they will "fit in", and often this is based on the surrounding culture as well. She also notes that not many companies have really strong programs for equalizing diversity. Businesses need to take stock of whether or not the diversity of their employees is represented at all levels of the organization and within all areas.

Work–Life Balance

In the past, Sharon admits her work quite outbalanced her life, but that the imbalance is slowly being corrected. At times work must be a priority, but at times family must be as well. Her recommendation is to "take your vacations and use every single day. Also make time for yourself and family every week, even if it is only part of a weekend. Remember that this time is as equally important as your work like!" In her free time, Sharon enjoys aerobic kickboxing, photography, and traveling.

Final Thoughts

Finally Sharon suggests that each of us decide whether we want to be a Jack-of-all-trades or a master of one? For Sharon, a satisfying career includes the flexibility to change the specific nature of what she is doing every three to five years. She likes learning about the different functions and areas where the business is active and has always tried to keep directed towards positions that will allow her an opportunity for change later. But that is not for everyone. Sharon has great respect and admiration for the technical experts that she relies on daily, who have spent a significant portion of their career in one area and understand those systems inside and out. Although that is a path she did not choose to follow, it is just as valid and can be equally satisfying.

Sharon also notes that you need to decide whether you prefer to be a single contributor or a group leader? Look for job characteristics that you think would make a job interesting for you. Do you prefer technical-based activities or negotiating? Also, carefully consider your preferred environment. The impact of the people and culture in the place where you work cannot be underestimated and should be taken into account when you look at where your career is and where you want to go, reminds Sharon.

Sharon's advice is to always be willing to change course. "Finally, despite all your planning and mentoring and self introspection, you may still find that your choices aren't bringing you the satisfaction you thought they would. If that should happen, it is simply time for a course correction and not a sign of failure!

Chapter 24

Tanya Travis, General Manager: Building Confidence

Jody A. Kocsis

Technology Manager, Engine Oils, Lubrizol Corporation, 29400 Lakeland Boulevard, Wickliffe, OH 44092

Life requires finding the middle ground. Tanya Travis couldn't think of any situation in life where you don't have to make choices and those choices have consequences. She has made many choices along the way of building her career, and they didn't all turn out rosy. Making informed choices is key.

After Tanya Travis graduated as the Salutatorian from East Technical High School in 1976, she initially went to Brandeis University, a liberal arts college in Massachusetts. She had a full-tuition scholarship and a living stipend and was majoring in Chemistry as a pre-med student. Tanya experienced not only college life but also a time of turmoil. Tanya remembers experiencing the racial tensions in the Boston area brought on by busing. African-American students received a great deal of guidance on areas of the city to avoid for their safety. She was accustomed to a predominantly African-American environment while growing up in Cleveland, Ohio. She was used to going where she wanted and in the past hadn't worried about her safety. Moving to Massachusetts and attending a university, which was at that time predominantly Caucasian and orthodox Jewish was a significant cultural change.

After her freshman year, Tanya decided she wanted to be closer to her family in Cleveland and that Chemical Engineering was more appealing to her. She transferred to Case Western Reserve University (CASE) in Cleveland, Ohio, where she graduated in 1981 with a B.S. degree in Chemical Engineering. In 1980 she married her husband, James.

Tanya Travis (Courtesy of Lubrizol Corporation.)

While at CASE, Tanya was very fortunate to participate in the Minority Engineering and Industrial Opportunity Program (MEIOP). This program focused on providing students with experience in industry as well as support for effectively navigating through an engineering curriculum. It also afforded her the opportunity to work in several engineering-related functions, as an intern, throughout her college career. She worked as an intern at several corporations: BF Goodrich, Standard Oil, and Calhio Chemicals. Tanya's intern experiences afforded her the opportunity to get a clear picture of what she wanted to do with her career. They really helped solidify her interest in engineering and guided her toward working in a manufacturing environment..

After graduating from CASE, Tanya started her career at the Lubrizol Corporation in Painesville, Ohio, one of their many manufacturing facilities. She has held almost every position possible in manufacturing, including process engineer (1981), section engineer (1981–1986), section superintendent (1987–1990), operations manager (1994–1997), manufacturing manager (1998–2000), and now general manager (2003–present). After her first managerial position at Painesville, she spent four years in Lubrizol's Corporate Operations Group, working as an operations manager for facilities in Southeast Asia and Latin America. She also had the opportunity to spend three years, (2000–2003) in Research and Development, as the Department Head for the Process Development Department. Of all the opportunities she has had at Lubrizol, her most memorable experiences were with the movement toward self-directed work teams and the Painesville Plant re-design initiative. These eye-opening experiences solidified in her mind the true value of Lubrizol's employees and how effective they can be with a full understanding and perception of their role in achieving a goal. In her current role as General Manager of a manufacturing facility, she oversees all the activities at the Painesville facility and ensures that operations take place in a safe, and environmentally responsible manner while meeting the demands of Lubrizol's customers and supporting the community.

Supportive Employer, Work Climate

Research is critical when determining a career and choosing a potential employer. Tanya is a firm believer that, in most cases, you spend more years of your life working than doing anything else. A satisfying career first and foremost has to be fun and personally satisfying. Doing something you enjoy just makes you want to do it more. Your work environment is very important, as you will be spending much of your time there.

When she first started working at Lubrizol, very few women in a manufacturing plant had technical roles. It took Tanya some effort to be recognized as a viable contributor. Many of her coworkers didn't think that a woman belonged in a manufacturing environment. Being a minority didn't help either. Also, back in the early 1980s, many women quit work once they started their families. Tanya remembers being uncomfortable telling anyone that she was expecting her first child. She had her fair share of the comments surrounding the need for women to be home raising their families. Establishing herself as a credible contributor was key.

Since that time, Tanya believes the organization is much more understanding of issues surrounding families in general. With the addition of more women who have managed to balance work and family life in technical roles, she believes there is less of a concern. She also believes organizations are becoming more accommodating. Although, Tanya says, it is still tough at times to balance work and family.

Balancing Work and Family

Tanya has been married for 24 years to James and they have three children ages: 13, 16 and 19. She was fortunate when her children were younger that her husband worked for an organization that had an on-site childcare facility. She realizes everyone doesn't have that option but it helped her family tremendously. Tanya believes that balancing priorities and setting some ground rules for your operations are key to making any situation work. Too much of anything is not good for you. Tanya also points out that working in an organization that was very supportive of the need for balance in your life is very helpful. A person has to be able to communicate what they need to support their success and work toward striking that balance.

Balancing work and life can be a challenge if you don't manage it. Tanya has worked hard to keep them separate. When she needs to focus on work, that is what she does, but she makes sure she gives herself some "life" time. Tanya does some work at home and sometimes works late, but she has made it her personal commitment to not make it a habit. It is hard to stop once you start. Balancing work and life was much harder when her children were younger, since they needed more of their mother. It is a lot easier now that they are older, but there are still times when she has to put them first.

Mother and Professional

Being both a mother and a professional has been probably the toughest challenge for Tanya. She absolutely had to make choices with regards to work, which took her away from home. She says, you balance that like you balance everything. There is always that little shred of doubt about whether you somehow have not given your child everything he or she needs to be successful in life. Tanya says that time will tell. She must say that her kids are great! They're happy and have done well in school. They have also found their own activities, which help them provide some balance to their own lives. Tanya says they don't take elaborate vacations because she and her husband believe in investing in their children, since they are our future. Tanya taught her children that their actions represent who they truly are. How you present yourself to others around you will have an impact on how others will treat you.

Mentoring: Past, Present, and Future

Tanya shared her thoughts about mentoring. Mentors must be willing to give freely of themselves to benefit others. Knowledge of the right roads to take and the pitfalls to avoid is important, as well as the ability to communicate and generate excitement in your activities. Students can seek out mentors by studying the behaviors and qualities of those they admire and wish to emulate. Having connections (i.e., networking) cannot start soon enough. For college-bound students, the parents of friends as well as acquaintances of parents and teachers might also be a good route to pursue. Once a candidate is found, look for an opportunity to get to know the individual and ask this person to be a mentor.

Tanya found one of her first mentors in high school. Tanya's interest in science was sparked early in her high school years, as a sophomore. She had a great chemistry teacher, Mary Nackley, who made chemistry fun and interesting. These are the teachers you will never forget, says Tanya. Additionally, in the early part of Tanya's career, she worked for some great guys! And yes, she means guys. She has never worked for a woman. In every case where the relationship was good, it was clear that she was accepted and that there was a genuine interest in her success.

Definition of Success

What took Tanya to where she is today is pure and simple: focus. Tanya focuses on achieving whatever she sets out to do. The abilities to communicate and to work effectively with others are important for building confidence. She has seen this first-hand by working for and with some great leaders. Tanya is not a person who started her career with a long list of goals that she was working to accomplish. Two things define success for her: personal happiness and the ability to see her contributions making a difference. She approaches every opportunity as if it will be her last.

Difficult Times

As with many organizations the need to downsize and reorganize has not escaped Tanya as a manager. She was involved with a 50% layoff of the plant workforce. In 2000 she was involved in reducing the plant workforce by approximately 200 employees. Tanya said this situation was one of the worst times in her career as a manager. These are folks that trusted you. It is not easy, and you share their pain. Tanya is aware of the impact on a person's livelihood. You give them the respect and support that you can, but there is no easy way to terminate one's employment. After the layoff, rebuilding the trust and relationships with the 200 remaining employees took time, but in this economic uncertainty everyone must move forward.

Career Advice

Tanya advises that a good chemist or engineer starting out on a career quest or looking for a change must be confident, level-headed and be driven to succeed. A chemist or engineer must love to jump in and solve a good problem. The curiosity of understanding how things operate or why they are the way they are is important. In this day and age, these professions are predominantly male, so self-confidence and knowing when to assert oneself are also pluses. Also, given the whole economic climate, one should always be prepared with an updated resumé and take one to two seminars a year, in order to continue some formal personal development. It pays to be prepared. You never know when an opportunity might come your way.

Chapter 25

Melissa V. Rewolinski, Director of Chemical Research and Development: A Career in Teamwork and Leadership

Rita Majerle

Hamline University, 1536 Hewitt Avenue, St. Paul, MN 55104

Teamwork is essential for project success in industry. This profile illustrates how one woman has successfully gone from the individual project focus of graduate school to the role of industrial group leader. It also shows that chemistry is definitely a people-oriented career.

Melissa Rewolinski is a director and section leader of the IMI (Isomers, Metabolites, and Impurities) Group within Chemical Research and Development at Pfizer Global Research and Development In La Jolla, California. She is responsible for supervising and training group members and is strongly involved in Pfizer's local and global recruiting efforts.

The vision of a lone scientist in a lab carefully guarding his results is a far cry from reality for today's scientist. Melissa adds that teamwork drives industry. "Industry would not be successful if there weren't common overall goals for their employees." Therefore, teamwork is a must. "All projects are approached by teams of people who have the success of the project as a common goal. However, within that, there is still plenty of opportunity for individual achievement. I think that the best companies to work for are going to provide opportunities in both areas. One important thing to note, however, is that when we are looking at candidates for positions within Pfizer, we have to see their capability for getting along with others (ability to work in a team environment) as well as scientific excellence. Scientific excellence alone will not get you a job."

Melissa V. Rewolinski (Courtesy of Jay Gould.)

Education and Early Career

Melissa was born in Georgia and grew up in Tennessee. Her father has a Ph.D. in history and studied theology and geography. Her mother holds a master's degree in health management. Although not the daughter of scientists, she grew up in a household where education was valued and knowledge for its own sake was important.

Her first experiences with honors high school chemistry made it clear to her that this would be her career pathway. She had a special project in which she looked up the procedure and then synthesized acetaminophen (Tylenol). That was it for her. She was excited that, "I could make something that someone could take and make their life better!" Although she briefly entertained the idea of studying chemical engineering, it wasn't the direction that Melissa wanted to go. "My favorite course was my first-year organic course," she stated. Organic chemistry, in particular synthetic organic chemistry, was the field she decided to pursue.

Melissa attended Rice University in Houston and obtained her Bachelor of Science degree in 1991, graduating Magna cum Laude. During her summer breaks as an undergraduate, she experienced industrial research first hand. As a sophomore she used optical and scanning microscopy to determine possible causes of polymer failure as a microscopy technician for Dow Chemical, in Freeport, Texas. The next summer, Melissa became a summer research chemist at Shell Development in Houston, synthesizing consumer chemicals and fuel additives.

She returned to Rice University after her graduation to pursue her doctoral studies under the guidance of Professor Marco A. Ciufolini. Although her thesis involved the applications of 'ene' reactions to total synthesis projects, Melissa learned even more important and practical information. She credits Ciufolini with being a strong mentor as well as a good advisor. "I wasn't sure if I wanted to pursue a career in academia or in industry," admits Melissa. Watching her mentor Ciufolini, she decided that although she enjoyed teaching, teaching was not as valued as proposal writing and obtaining grants in some institutions. She notes, "I didn't want to miss doing the science."

From Houston, Melissa went north to join Pharmacia and Upjohn in Kalamazoo, Michigan, for an industrial postdoctoral appointment. As a member of the medicinal chemistry group, under the direction of Dr. Steve Tanis, she worked on the synthesis of biologically important alkaloids. It was an interesting experience being a member of a multidisciplinary team that studied mechanism-based disease targets. Melissa commented that, although taking a postdoctoral position is not an ideal way to get a position in the larger pharmaceutical firms, the economy at the time dictated her choice and in the long run it worked for her.

Work and Life

Compromise is a necessity when pursuing a career and nurturing a growing family. "Balance is key," says Melissa. Her family includes an active three-year-old son and another child on the way. When asked if she can "do it all", Melissa replied that compromise is an important part of being able to juggle work and family. Each individual, says Melissa, determines the definition of success. "I feel that I'm successful because I am making a difference. It's important to me that I make a difference in both my work and my family", she adds. When Melissa isn't working, she enjoys reading, jogging, traveling, sleeping, spending time with her family, and going to the beach.

Professional Development

Mentoring has made a big impact in Melissa's life and career. Although she is not a part of a formal mentoring relationship in her current position at Pfizer, she credits Ciufolini and Tanis with being strong and helpful advisors in her career. Additionally, Melissa has found that networking is a very important part of career development. These connections inside Pfizer and throughout the profession are always valuable.

Although time spent on professional development is at a premium, Melissa makes an effort to attend professional meetings such as a Gordon Research Conference on a yearly basis. Pfizer offers in-house courses and, when asked about a favorite, she replied that a course on organometallic chemistry by Jacobsen and Buchwald was right up there. Management short courses have also been helpful in her career path as she notes that she is "no longer a bench chemist" but has moved into a supervisory role. She also makes a point of updating her resumé at least yearly. As a recruiter of Pfizer, she is in tune with the job market and recommends keeping an eye on the economy.

When asked what was her favorite book, Melissa, with her full-time career and growing family, replied that _Who Moved My Cheese_, by Spencer Johnson, has been inspirational. It is a parable in which the characters find that consciously choosing to adapt allows you to take advantage of new possibilities. In a rapidly changing industry such as pharmaceuticals, adapting is a necessity. On a more philosophical note, Melissa says that a key part to growth is to take chances. A phrase that is posted near her desk sums it all up: "What would you do if you weren't afraid?"

Final Thoughts

When asked what made a position in industry challenging and rewarding, Melissa replied, "It is always changing. There is a constant need to do things differently. Thinking outside the box in such areas as process technology for example and applying and using new methods like the use of enzymes and such is very exciting." Her advice to young women starting out is to be pro-active in your career development, to always think of where you want to be and then match that up with your plan on how to get there. "Be strong," she says. "Don't feel like you have to accept anything less than what is right and fair for you or for others around you."

Chapter 26

Margaret A. (Lissa) Dulany, Chemical Consultant and Writer: Chemistry and People

Amber S. Hinkle[1] and Margaret A. (Lissa) Dulany[2]

[1]Bayer Material Science, 8500 West Bay Road, MS–18, Baytown, TX 77520
[2]MADesign, Inc., 201A 5th Street, NE, Atlanta, GA 30308

To those who know and love her, Lissa Dulany is an outgoing and charismatic woman with a very sharp mind. Thus, it is not surprising that her career path has consistently led through positions where she combines chemistry and personal interactions very successfully and joyfully.

Lissa Dulany grew up in Maryland, always living within sight of the Severn River. From an early age, naturally, she wanted to become a marine biologist or oceanographer. She had a lot of fun participating in several science fairs during high school, and her projects always had a marine focus. During this time, Lissa also won a trip to Florida from the U.S. Navy, which she believes was both socially and academically important to her future development. Her scientific shift towards chemistry began during her senior year in high school, when she says, "the advanced chemistry course, taught by a feisty older woman, Ms. Virginia Siler, was both a ton of fun and very interesting." Lissa's subsequent journey in chemistry has been interesting in and of itself.

Education and Career Path

Lissa pursued her undergraduate education at the University of Virginia, where she completed a dual major in Religious Studies and Chemistry, affectionately known as "alchemy". After graduation, Lissa launched off in one

Margaret A. (Lissa) Dulany (Courtesy of Margaret A. (Lissa) Dulany.)

of those directions by taking a position as a youth minister in Atlanta, Georgia; employed by an organization with whom she had learned much as a teenager herself. While immersed in teenage culture and counseling kids through numerous opportunities and crises, Lissa also worked on a master's degree in community counseling at Georgia State University. She notes that those times were very personally rewarding, but after four years Lissa chose to pursue her other major and was accepted into the doctoral program in chemistry at Emory University.

Her graduate work was centered in so-called "organized assemblies" of molecules, from host–guest dynamics in modified cyclodextrins to noninvasive structural investigations of micelles, phospolipid bilayers, and biological membranes. Much of this physical organic chemistry work was accomplished by the synthesis of C-13 labeled molecules that were then studied by high-field NMR. Lissa almost accepted a postdoctoral position in a very similar investigative area but chose instead a position as a product development chemist in the chemical division of a large forest products corporation, Georgia-Pacific. During her 12-year career at Georgia Pacific, Lissa enjoyed immensely the interaction with other research personnel and numerous customers, with whom many technical projects were successfully completed. She also had the opportunity to work with many manufacturing sites, as new products were developed and brought to market. The businesses served during this time were the paper industry (mainly wet and dry strength additives for paper and paperboard) and insulation/foundry/lamination resins for numerous industrial applications.

Due to her love of the technical/commercial interface, Lissa next accepted a position as the Technical Service and Development Manager for the U.S. operations of the Belgian chemical/pharmaceutical company UCB. During the next 6-and-a-half years she was fortunate to hold several similar positions in both research and technical service, in three very different industrial chemical businesses, as it was company practice to cross-train the technical management staff. Lissa thrived on the customer technical interactions involved in developing specialty coating solutions via radiation-curable resins and thermosetting powder resins. During a more recent re-organization, her background was utilized in the plastic packaging films business (polypropylene and cellophane). This business was subsequently sold, and Lissa now has the exciting opportunity to develop yet another unique creative expression for herself, utilizing the entirety of her industrial chemistry career. She knows that this next phase will be more entrepreneurial and will involve communicating chemistry to children, among others. Lissa invites us to "stay tuned" to see what new peaks she reaches with her newly started consulting career.

Opportunity

One of the veins running through Lissa's career is "opportunity thinking". By this she means the consistent practice of seeing opportunities everywhere. Lissa says she is by nature an expansively creative thinker; "I readily visualize connections between seemingly disparate items." Strategically, this quality is very useful, but Lissa has felt a bit "out in front" at times, waiting for others to see the same opportunities through their own lenses.

Lissa noted that one of the other advantages she had in her career is that often she was the only woman in the room, in the manufacturing plant (except for administrative personnel), on the business team, at the customer's facility, etc; which automatically made her memorable. Also she always asks questions, and especially with customers and manufacturing environments her interest is evident. Thus, the people with whom she works are happy to help her understand whatever she is asking. Lissa especially does this if she thinks she already knows how something works, since it fills in the gaps of her understanding. One of the disadvantages, she says, is that everyone could remember her name, which made for some awkwardness if she was not as prepared as she should have been.

Success

Lissa defines success roughly as "being able to do what I want, where and when I want, with whomever I want—this sense of freedom is at the very heart of my happiness." Lissa also notes that her definition of success encompasses the joy of learning new technologies, solving technical challenges, meeting new people, and traveling to new places, all of which she has done many times so far during her industrial chemistry career. However, she has made some compromise over the years, sort of a dance in time and space, as work challenges alternatively rose and dissipated in a living wave, she admits. Lissa says that her so-called "work–life" balance has improved in recent years, as she has learned to selectively protect her free time. For example, she has been able to complete several years of classes in her spiritual community, deepening her ability to sense what is really important. Also, vacations for her often have been adventurous and steeped in service, including two trips into Peru and Machu Pichu, a magical place. Lissa can also be found sometimes with sand between her toes on the shore of the Gulf of Mexico.

Networking

Informal mentoring, where she sought out individuals who knew something she wanted to learn, has been a factor in Lissa's success, along with lots of curiosity and the confidence to participate on and/or lead new projects and teams. Neither of her companies had any formal mentoring programs, but she found her own mentors quite readily. Lissa also notes that her professional networks are broad, both within the corporations where she was employed and within the American Chemical Society and other professional or trade organizations. "Networks must be both functionally broad and hierarchically diverse," she reminds us. In her career Lissa has also been fortunate to develop global networks, although these have been predominantly technical and most recently social.

When discussing success, Lissa could not say enough about the benefits from volunteering in one's professional or civic organizations. Not only does it provide immense satisfaction, one can also learn additional skills that can then be used within one's primary career if desired; sort of on-the-job training but off-the-job. In addition, the networks and connections are invaluable, whether it's brainstorming a technical challenge or seeking a new employee or employer. Lissa herself has spent much of her personal time and resources providing service within organizations such as the American Chemical Society.

Final Thoughts

Lissa left us with some profound advice, "What trait makes the best professional, chemist, manager, spouse, parent, or friend—the ability to communicate clearly. There is no substitute—this includes asking provocative questions, giving concise answers, and listening intently throughout. It is critically important for me to sense my audience always, in order to communicate cleanly and appropriately."

In making choices throughout your career, Lissa recommends "follow your heart and do what you love! This is both more healthy and fun and will ultimately lead to your greater success and happiness and to a greater contribution to those around you, personally and professionally." Lissa has certainly demonstrated the wisdom of this philosophy in her own life!

Chapter 27

Shirlyn Cummings, a Human Resources Perspective: Leaders for a Global Market

Amber S. Hinkle

[1]Bayer Material Science, 8500 West Bay Road, MS–18, Baytown, TX 77520

As part of a study on successful women chemists in industry, the manner in which industrial companies are dealing with retention and development issues must be considered. Shirlyn Cummings is Director of Human Resources for Bayer's largest U.S. manufacturing site. She has an interesting perspective on these questions.

Shirlyn Cummings, Director of Human Resources, for Bayer's Baytown, Texas, facility, was interviewed in regards to Bayer's programs for retaining and developing women, scientists and others. As Shirlyn so aptly puts it, "Over the years, more women are increasingly represented in a variety of positions across Bayer and the Baytown site. This brings an exciting new perspective to things. The benefits of having diverse viewpoints on the same problem have garnered much greater realization and acceptance. Hiring is talent-oriented versus gender-based; managers simply want 'the best'." However, Shirlyn is also aware that human resources departments need to pay special attention to keeping 'the best' with the company. She notes that Bayer has come a long way in working to meet special needs while also balancing this objective with good working conditions for all employees.

As evidence of this progress, Bayer won the coveted Catalyst Award in 2002 for programs related to integration and retention of women and minorities. This recognition was due to several winning strategies, according to Shirlyn. Two important areas that Bayer has focused on are mentoring and diversity training.

Shirlyn Cummings (Courtesy of Shirlyn Cummings.)

Mentoring

Shirlyn describes various mentoring programs within Bayer including one-on-one mentoring and new-hire mentoring.   In the one-on-one mentoring programs, the main objectives are to focus on the mentee's growth and development and to provide equal partnership with two-way learning. One such program pairs mentors from senior management with high-potential employees, while another program asks for volunteers to be mentors and mentees and creates partnerships based on the mix of employees who volunteer. These mentoring partnerships typically last for at least one year, although they often continue beyond the formal boundaries of the mentoring program. When these partnerships are created, attention is paid to the background, experience, race, and gender of each participant in an effort to maximize the diversity within the organization. These mentoring programs are overseen by the Office of Diversity and individual site Human Resources departments, who aid in the process by matching mentoring pairs, providing education on how to mentor, checking in with the partners periodically, and holding group sessions to share best practices and celebrate mentoring in general. In the one-on-one mentoring program, employees are selected by their business units to participate in as part of their career development paths.

Of course, Shirlyn says, the first mentoring program that employees might be exposed to at Bayer is the new-hire mentoring program. This program is available at some of the Bayer sites, and it pairs a senior manager with a new employee. The objectives of this program are to integrate the mentee into the business quickly; to facilitate learning, growth, and systematic development of desired skills; and to assist the employee in developing a professional identity that in turn enhances their competence and confidence. Mentoring has played a key role in the development of many successful women chemists, thus having the opportunity to participate in formal mentoring programs is invaluable.

Finally, Shirlyn mentions that "good mentors listen well and are confidence builders. They provide subtle guidance but don't shield you from opportunities or mistakes, and they share their wisdom without feeling threatened. Mentors relish and celebrate your successes!"

Diversity Programs

Bayer monitors its inclusion programs with a pyramid strategy, explains Shirlyn. Bayer has an overarching Diversity Advisory Council (BDAC), which is fed from the broader-based Diversity Leadership Forum, which in turn is fed by the Diversity Councils located at each Bayer site. The BDAC consists of people who can affect inclusion, from line to senior managers across the company, according to Shirlyn. This group advises the individual diversity councils in developing initiatives, as well as tracking and reporting progress and best practices. These diversity initiatives at Bayer include a two-day diversity journey training program for supervisors throughout the company and a one-day training for all other employees. During this training, Shirlyn explains, employees have an opportunity to discuss diversity issues and how they can promote diversity within their own groups. Bayer also has a diversity Web site and video library, where employees can find resources to assist them.

Besides general diversity training and information, Shirlyn says that specific programs can ensure diversity throughout the organization. One is the Associate Development Program, whose main objectives were to increase the ratio of females and minorities in internal, high-potential candidate pools and develop leadership talent for the future. In this program, targeted candidates are assigned a sponsor and a mentor, with whom they spend 24 to 36 months in rotational assignments and receive a specific developmental plan. Of the participants in this program, 57% are women. Shirlyn shared the following results of this and other similar programs:

- Percent of women at the executive level increased from 2.6% in 1998 to 6.7% in 2001.
- Percent of female Vice Presidents increased from 8.8% in 1998 to 12.8% in 2001.
- Percent of women Directors rose from 13.6% in 1998 to 21.6% in 2001.

Another program that Bayer prides itself in is the Delegate Career Development Program. According to Shirlyn, this program is designed to share the diversity of different cultures by sending employees on foreign assignments within Bayer. Not only do delegates from the United States have opportunities to work in locations in Europe, Asia, and South America, but delegates from those areas come to the United States as well. Preparation for this program includes intensive language training, orientation to the new culture, and assignment of a senior management mentor. These assignments typically last from two to three years and are key in career development within Bayer. Between 1997 and 2001, 20 women participated in international assignments through this program.

Local Initiatives

Bayer's Baytown, Texas, site is near and dear to Shirlyn's heart. She has been an active advocate for women at this site over the past 19 years. Thus she wanted to share some of the progress made at this site in particular. Since 1996, the number of women managers at the site has doubled and the percentage of women executives has gone from 0 to 31%. This is outstanding for a chemical plant, according to Shirlyn, who also participates in study groups along the Gulf Coast industrial corridor.

The question that Shirlyn and her colleagues at the Baytown site ask themselves is "What has made the single biggest difference to the advancement of women at Baytown or, more importantly, what is expected to make the single biggest difference for women at the Baytown plant in the future?" Shirlyn believes that the leadership philosophy adopted by the Baytown plant makes all the difference. Leaders are held to these standard practices and expect their employees to follow them as well:

- Inspire a shared vision.
- Model the way.
- Encourage the heart.
- Enable others to act.
- Challenge the process.

All employees are evaluated on these behaviors, and senior management monitors an overall site leadership index. Over the past four years, this leadership index has risen to 95% of the "Best in Industry" level. Shirlyn notes that emphasizing excellent leadership helps all employees, including women.

Thus, her specific advice to corporate decision makers is "to promote strong leadership, because strong leaders value people and valued employees want to stay." Shirlyn also recommends that leaders strive to give their employees a sense of purpose and, above all, be flexible. The old way of doing things was to treat all employees exactly the same, the better way is to treat each individual with respect and meet her specific needs, to the extent possible. Shirlyn believes that these practices will result in win-win alternatives for all employees.

Working Mothers

Other factors encourage women to stay in the work force while balancing work with their homelife. Shirlyn mentioned that Bayer is listed on the Top 100 List of Best Companies by *Working Mother* magazine. Bayer made the list in 2002 and again in 2004 in how its work–life programs stacked up compared with other companies across the United States. The types of services, for working mothers and others, that Bayer provides at some or all of its sites include:

- Offering on-site or near-site childcare
- Sponsorship of sick childcare
- Providing childcare resource and referral services
- Sponsoring before- and/or after-school childcare
- Offering compressed workweeks
- Offering flex schedules
- Offering job-sharing
- Offering telecommuting
- Providing resource and referral services for eldercare

- Having lactation programs for nursing mothers
- Offering an employee assistance program

Shirlyn notes that it is smart business to provide these kinds of services for all employees. The economy may not have been conducive to investing money in these programs lately, but the payoff is for the long run. The forecast for the future looks very dim in regards to the shrinking talent pool in the sciences. Companies with sustainable programs for development and retention of talented employees will continue to be the work places of choice.

On a Personal Note

Shirlyn is a successful woman in her own rite. She has worked for Bayer for 19 years and is one of the highest-ranking women at Bayer's Baytown site. She shared with us her personal success story, as well as her advice from a human resources viewpoint. Shirlyn says that integrative thinking and the ability to see how the pieces fit together got her to where she is today. She incorporates what others might consider passing comments into the big picture and strives to match people's skills and personality with the right job. As to Shirlyn's favorite pastime, it is her 3 children, ages 10, 12, and 13. They keep her sane, and she believes that balancing family with work makes her whole and enables her to be better at everything she does.

Success and Balance

Shirlyn believes you are successful when you have balance and happiness, whether at work, at home, or both. She says, "You should always have a mission in life, allow yourself to get excited about changes, and never settle for boredom. A lack of success is when you just 'put your time in'. Working, like parenting, offers great opportunities to learn and grow as a person. 'Treading water' is okay, while you get your bearings, but to really make progress you have to swim with strong, sure strokes."

In order to be successful, Shirlyn recommends that women invest in themselves by developing their leadership skills. She also says that women should be flexible, find a purpose, and "just do it...the best successes are rarely assigned." She notes that you should always project confidence, and then figure out what you do not know afterwards. Also, if you value people and relationships, the rest will fall into place, Shirlyn believes. However, she also cautions that you must show good judgment in whom you share your self-doubts or intimate thoughts with at work.

Shirlyn's final thoughts surround work–life balance issues. "At work, delegate ahead of time; do not unnecessarily make yourself the focal point for everything. Hire and utilize good people. Be able to receive help, and give it in return. Don't push others away. At home, take time for self-reflection, exercise to relieve stress, and find ways to leave work at work, which I still struggle to do. In everything, don't sacrifice quality for volume. Take vacations!! Also every day find 20–60 minutes of quiet time just for you. Use this time to think through what's required for the rest of the day and adjust your attitude to be confident, in-control, and prepared."

Chapter 28

Lessons Learned

Arlene A. Garrison

Office of Research, University of Tennessee, 409 Andy Holt Tower, Knoxville, TN 37996

Common threads were identified in many of the interviews for this book. Some were validated by larger studies. This chapter addresses some of the tools for success identified by a number of women.

Family support and mentors were the most often recognized tools for success in the interviews for this book. The women advanced in their careers during an important period of history where workplace diversity was strongly emphasized. Traditionally underrepresented groups, including women and ethic minorities, were receiving degrees in significant numbers and the pipeline problems of previous generations were disappearing. The successful women interviewed for this book often mentioned the importance of participation in professional associations.

Mentoring

Role models and mentors have long been recognized as important tools for career advancement[1]. Mentors normally hold higher-level positions and provide suggestions and advice to women seeking a successful career. Many of the women interviewed for this book attribute their success to a relationship that provided support and information about unwritten rules and processes. Due to the demographics during their careers, many of the women in this book acknowledge that their primary mentors have been men. Such male mentors can serve as role models for a technical discipline. The women also acknowledge other role models and support networks to sustain the need for confidants and suggestions about family issues. Many of the women in this book serve as role models and mentors to a number of younger women at the early stages of their careers. In many situations, this early support has been found to be critical to the retention of talented young women in science careers.

Workforce Diversity

Many organizations and researchers have discussed the importance of greater involvement of traditionally under-represented groups in science and engineering careers in the United States today. Largely due to population shifts, an increase in the diversity of the workforce is happening in all fields. A recent report by the National Science Board, "The Science and Engineering Workforce: Realizing America's Potential"[2], provides data supporting the need for a greater supply of domestic graduates in science and engineering. A variety of viewpoints is critical to the success of multidisciplinary team projects. Such projects are very common in industry today. Some specific technology careers have achieved success in increasing participation by groups such as women and racial and ethnic minorities. The lack of diversity must be addressed in order to maintain viability of industries in the United States that depend upon scientific and engineering expertise, particularly as the current workforce ages.

The advancement of women in science and engineering has been studied extensively by a number of researchers. Numerous publications address the challenges for women who aspire to manage in a technology field. In "Athena Unbound"[3], Henry Etzkowitz discusses the myth of the pipeline. The percentage of Ph.D.'s awarded to women has risen dramatically over the past few decades, from less than 1% of the engineering doctorates in the 1960s to 11% in the 1990s. The percentage of chemistry female Ph.D.'s has increased from 7 to 27% in the same time period. Despite the increase in participation and graduation, women continue to be under-represented in leadership positions. Etzkowitz discusses the many challenges to women who wish to advance in these fields with particular emphasis on the system of "social capital", the relationships and networks critical to success in highly interactive technical fields. Women are often excluded from critical interactions and social networks that are key to advancement in technology disciplines. This may be true, but buy-in from management is also necessary from the top down, as cited in this book by several interviewees. Etzkowitz also discusses the importance of having a critical mass of women in an organization to implement sustainable change. His study is particularly interesting as it analyzes the issue on an international basis.

Table 1. Percentage of women in science and engineering fields (after Etzkovitz[3]).

Increased Science and Engineering Ph.D.'s, percent female, 1966–1997

	1960s	1970s	1980s	1990s
Chemistry	7	10	19	27
Mathematics	5	10	15	19
Physics	2	4	8	12
Engineering	<1	1	6	11

Virginia Valian's book _Why So Slow? The Advancement of Women_[4], provides insight as to the reasons for the limited number of women in high-level positions. She has published numerous studies and has developed a Website[5]. Her work has focused on the gender schemas that lead to barriers to women's success, particularly in nontraditional fields such as chemistry and engineering. Gender schemas are expectations or ideas, more neutral than stereotypes, which permeate a culture and are carried by both men and women. These ideas are often basic and are derived naturally over many years of experience. For example, the common expectation that women are shorter than men is normally correct. Assumptions about leadership skills that reflect a gender basis are also likely to be correct in a technical world. When entering a conference room, meeting participants often assume the man seated at the end of the table will lead the meeting and that the woman seated to the side is a support person. This interpretation may be valid or can reflect different personality styles of individuals. Schemas, although not ill-intended, can lead to barriers to success for women and other under-represented groups in science and engineering.

Valian presents many conclusions from her extensive research. First, progress has been made. Men and women now start their careers at roughly equal pay and rank. Second, problems remain in that men advance more quickly than women. Finally, she finds that the problem is not unique to science and engineering. It exists in many disciplines, not just technical fields. Some of the interviewees in this book also mentioned that women tend not to broadcast their successes as loudly as men, if at all.

Professional Associations

Many of the interviews for this book mentioned the importance of professional associations. The organizations often provided a venue for growth in leadership skills. In cases where women were excluded from the important networks in their companies, professional associations provided an alternate venue for information and networking. The management skills learned through volunteer work in professional associations were often identified as tools to learn how to manage a diverse, multidisciplinary work force. Targeted organizations, such as the American Chemical Society Women Chemist's Committee, provide a venue to obtain the necessary skills.

Professional associations provide a number of additional benefits to participants. Many members consider only the financial benefits, such as discounts for conferences and journals and cost-effective insurance due to the large pool of participants. A typical career now involves several jobs. Professional associations are an excellent source of short-courses and other life-long learning. They also provide a network for job searching. Some scientists find friendships that last for decades through committee work. The women interviewed for this book point to mentors as critical to their career success. Some of these women identified mentors in their company, while others have identified mentors with similar interests through their professional organizations. Professional associations serve as a voice for groups of individuals with common interests and problems, especially with elected officials.

Conclusions

Professional associations provide an excellent opportunity to build leadership and management skills. Diversity of individual scientists and engineers is important to building a strong industrial team with the necessary problem-solving skills. Language and cultural barriers arise in many projects due to differences in age, gender, and ethnicity of the project participants. The team goal can only be achieved if all the participants communicate. The understanding of gender schemas can be very helpful in identifying the mechanisms of miscommunication.

Demographics in the chemical sciences continue to change. Groups that have been historically underrepresented are now participating in higher numbers.

Many educational programs are in place to provide access to women and minorities. Companies are measuring the positive impact of diversity and are demanding a diverse pool of graduates from the universities. Today's generation has a better awareness of the need for balance in personal interests and career growth.

Chemists and chemical engineers today face many challenges. Women in the chemical sciences often speak of the need to find a balance between competing demands. Some studies have indicated that women select lower stress, non-management positions as a means to balance their personal and work lives. Although more studies are needed to obtain a definitive answer to this question, the possibility exists that only women are perceived to make such choices, and in fact gender schemas are at the heart of this interpretation. Managers must be supportive of workers on their teams and be careful to avoid assumptions regarding goals and aspirations. Women with an interest in advancement should find opportunities for leadership training and must make their intentions clear to co-workers. Professional associations provide an excellent venue for growth and building networks.

---

1. *The Woman Scientist*, Clarice M. Yentsch and Carl J. Sindermann, Plenum Press, New York, 1992.
2. "The Science and Engineering Workforce: Realizing America's Potential", National Science Board, NSB 03-69. http://www.nsf.gov/nsb/documents/2003/nsb0369/start.htm
3. *Athena Unbound*, Henry Etzkowitz, Carol Kemelgor, and Brian Uzzi, Cambridge University Press, Cambridge, UK, 2000.
4. *Why So Slow? The Advancement of Women*, Virginia Valian, The MIT Press, Cambridge, Massachusetts, 1999.
5. http://www.hunter.cuny.edu/gendertutorial/

Meet the Authors

Jody A. Kocsis

Technology Manager, Engine Oils, Lubrizol Corporation,
29400 Lakeland Boulevard, Wickliffe, OH 44092

As with any good book it is nice to know a little bit about the authors. Below are descriptions of the authors and editors for this book.

Collecting and sharing interviews of successful women chemists has been an on-going project for the American Chemical Society's (ACS) Women Chemists Committee (WCC) since 1 998. T hus, t he w omen c ited b elow w ho m ade t his book possible are all current or previous members of the WCC. Each of them has d onated c onsiderable p ersonal t ime t o c onduct and write these interviews, which means a lot because personal time is a very valuable commodity. Additionally, a ll of these women are successful chemists themselves, and they believe strongly in encouraging women into the sciences and assisting them in their careers.

Women Chemists Committee

The Women Chemists Committee celebrated its 75th anniversary in 2002. It has numerous programs to support women in the chemical sciences. The committee of volunteers is organized in subcommittees to support its mission to be leaders in attracting, developing, and promoting women in the chemical sciences. Various projects are described in the newsletters and are listed below.

Authors: (back row from left to right), Anne Leslie, Ellen Keiter, Shannon Davis, Rita Majerle, Arlene Garrison, and Jacqueline Erickson. Editors: (front row from left to right), Jody Kocsis and Amber Hinkle. (Courtesy of Doug Hinkle.)

- Women Chemists Newsletter
- Canvassing Committee for ACS Garvan-Olin Medal National Award
- Active Website
- WCC On-line Mentoring Program
- Successful Women Series
- Women in Industry Breakfast at the national ACS meetings
- Women Chemists Reception and Luncheon at the national ACS meetings
- Sponsored and co-sponsored symposia

- Regional meeting activities
- WCC Travel Awards
- WCC Overcoming Challenges Award

The WCC travel awards program receives funding from several sources, particularly Eli Lilly & Co., which has been providing significant funding since 1989. Awards are presented twice a year to women presenting research for the first time at a major scientific meeting. Grant amounts are limited, and the educational institutions where the women are enrolled are encouraged to provide additional assistance. Awards can support travel to any appropriate technical meeting. The students who present at the ACS national meetings are recognized at the Women Chemists luncheon and participate in an additional poster session prior to the luncheon.

The WCC Overcoming Challenges Award was first presented in 2000. The award recognizes a female undergraduate student who overcame hardships to pursue a chemistry degree. Applications are solicited once a year. The student receives a cash award and participates in an ACS national meeting during the year of the award.

Members of the WCC have been very active in the ACS program called PROGRESS. The acronym represents a range of activities to support Partnerships, Reflection, Openness, Grants, Resources, Education, Site Visits, and Successes. The program was created as a result of an ACS Presidential-Board Task Force appointed in August 2000 to examine and make recommendations on issues related to women in the chemical professions. Four WCC committee members serve on the PROGRESS steering committee.

Shannon Davis

L. Shannon Davis graduated with a B.S. degree in chemistry from Georgia Southern College in 1984. She went on to earn a Ph.D. in Inorganic Chemistry from the University of Florida in 1988 and has worked in industry ever since. While employed by Monsanto and Solutia, her career has spanned bench chemistry, technology management, and commercial development. Today Shannon is currently the leader of the process research and development group for nylon intermediates. She is also a subcommittee Chair for the American Chemical Society's Women Chemists Committee and was honored to be interviewed for this book, as well as a contributing author.

Lissa Dulany

Lissa Dulany pursued her undergraduate education at the University of Virginia, where she completed a dual major in Religious Studies and Chemistry. After graduation, Lissa worked as a youth minister in Atlanta, Georgia, and earned a master's degree in community counseling at Georgia State University. She then went on to earn a Ph.D. in chemistry at Emory University. Lissa has worked in the chemical industry for almost 20 years now, first at Georgia Pacific and then with UCB. During this time, her responsibilities included product development, research, and technical service, as well as managerial and customer interface roles. Today Lissa uses her industrial chemistry background in new and innovative ways as a consultant and writer. Lissa is also very involved in the American Chemical Society and is a past member of the Women Chemists Committee.

Jacqueline Erickson

Jacqueline Erickson is a Senior Analytical Scientist at GlaxoSmithKline Consumer Healthcare R&D in New Jersey, where she has been employed since 1988. She holds a B.S. degree in Chemistry from the University of Delaware and an M.S. degree in Chemistry from Rutgers University-Newark. She is actively involved in ACS as a Councilor for the North Jersey Section and as an associate member of the Women Chemists Committee. Jackie participated in this book because she believes it provides excellent role models and demonstrates a wide variety of career opportunities for those in the chemical sciences.

Arlene A. Garrison

Arlene A. Garrison is Assistant Vice President for Research at the University of Tennessee (UT). Prior to her current role, she held numerous positions with UT, including Director of MCEC, an Industry/University Cooperative Research Center and Licensing Executive with the University of Tennessee Research Foundation, the agency that markets faculty inventions. She holds a B.S. in Electrical Engineering and a Ph.D. in Analytical Chemistry, both from the University of Tennessee. She is Councilor for the East Tennessee Section of ACS and is a member of the ACS Women Chemists Committee and served on the ACS Presidential Task Force on Women in the Chemical

Profession. Arlene has published numerous technical papers and has lectured at many universities and conferences throughout the world. In her local community, Arlene is on the Board of the Public Building Authority and the Board of the Southern Appalachian Science and Engineering Fair. In recognition of her volunteer work in science outreach to pre-college students, Arlene was one of the 10,000 Olympic Torch Bearers as the torch moved to the 1996 Olympic Games in Atlanta.

Amber S. Hinkle

Amber S. Hinkle is Director of the Quality Department for polycarbonate manufacturing at Bayer's Baytown, Texas, facility. Prior to her current role, she performed numerous functions for Bayer in both polycarbonate and over-the-counter medications manufacturing, including process chemistry, automated test method development, and lab management. She holds a B.S. degree in Chemistry from the University of Utah and a Ph.D. in Organometallic Chemistry from the University of Washington. She is a subcommittee Chair within the ACS WCC, is past Chair of the St. Joseph Valley ACS local section, and has served on the ACS Presidential Task Force on Women in the Chemical Profession. Amber has authored several technical publications and holds one patent for previous work. She has also spoken at the local and national level on her technical work, as well as on various women's issues. Amber served as an author and primary editor for this book because she has a passion for women's issues.

Ellen A. Keiter

Ellen A. Keiter is a professor in the Department of Chemistry at Eastern Illinois University (EIU). She earned a bachelor's degree in chemistry from Augsburg College and holds a Ph.D. in inorganic chemistry from the University of Illinois at Urbana-Champaign. She joined the Department of Chemistry at EIU in 1977 and, beginning in 1994, served for nine years as Department Chair. She is Councilor for the East Central Illinois Section of ACS and is an associate member of the ACS WCC. Author of numerous scientific articles, Ellen has also written, with co-authors J. E. Huheey and R. L. Keiter, the fourth edition of the textbook *Inorganic Chemistry: Principles of Structure and Reactivity*.

Jody A. Kocsis

Jody A. Kocsis has been a technology manager in the Engine Oil Product Development Group at The Lubrizol Corporation headquartered in Cleveland, Ohio, since 1989. She received a B.S. in Chemistry from Notre Dame College of Ohio, and holds 5 U.S. patents and 10 foreign patents for inventing new motor oil additives. She is a co-founder of Women in Lubrizol Leadership and a nominated member of the ACS WCC. Jody authored and edited this book because she views mentoring as a priority and wanted to assist others in finding their career potential in science.

Rita S. Majerle

Rita S. Majerle is an Associate Professor of Chemistry at Hamline University in St. Paul, Minnesota. She holds a Ph.D. in chemistry from the University of Minnesota where she trained in synthetic organic chemistry. As a member of the WCC, she has organized the successful "Women In Organic Synthesis" symposiums held at the ACS national meetings. Through her position, Rita has been an active mentor of young people in the fields of chemical and biological sciences.

Elizabeth Piocos

Elizabeth Piocos holds a Ph.D. and an M.S. in Physical Chemistry. She has had postdoctoral fellowships at the U.S. Department of Energy's Argonne National Laboratory and the U.S. Environmental Protection Agency. She is currently a senior scientist at the Procter and Gamble Company where she has been working in the Beauty Care and Feminine Care business unit. Beth's passion is in innovation: She believes in empowering women to be smart consumers and to be great innovators/inventors. She firmly believes in mentoring as one of the important tools to achieve these goals.

Frankie Wood-Black

Frankie Wood-Black is the Director of Business Services for Downstream Technology for ConocoPhillips. She is currently located in Ponca City, Oklahoma. Frankie has been with ConocoPhillips for 15 years and has held a number of different roles: bench researcher, environmental specialist, quality control manager, and technical marketing. She holds a B.S. in Physics from Central State University (now the University of Central Oklahoma), a Ph.D. in Physics from Oklahoma State University, and an M.B.A. from Regis University.

In addition to her business roles, she has been very active within professional societies. Leadership roles that she has held include Local Section Chair of ACS for the Northeast Oklahoma Section, the Salt Lake Section, and the North Central Oklahoma Section; Division Chair for the Chemical Health and SafetyDivision; and Chair of the WCC. Chemical education is one of her passions and as such has been very active with National Chemistry Week and continues to work with local elementary schools by visiting classrooms each month

Frankie is a contributing author to the *Journal of Chemical Health and Safety*, writing a bimonthly column. She was the co-author of *Emergency Preparedness—a Primer for Chemists* and an editor for the recently released Chemical Sciences Roundtable Workshop report on "Water and Sustainable Development". In addition to these writing activities, she has a number of technical papers and presentations.

Indexes

Author Index

Davis, L. Shannon, 1, 21
Dulany, Margaret A. (Lissa), 159
Erickson, Jacqueline, 17, 69, 79, 117
Garrison, Arlene A., 11, 37, 63, 135, 173
Hinkle, Amber S., 21, 75, 87, 139, 159, 165

Keiter, Ellen A., 31, 43, 111, 123
Kocsis, Jody A., 57, 93, 105, 147, 179
Majerle, Rita S., 51, 101, 153
Piocos, Elizabeth, 79, 101, 111
Wood-Black, Frankie, 1, 129

Subject Index

A

Academic careers, women, 4, 7
Advice
  Artman, Diane M., 108–109
  Barbour, Rachael L., 68
  Bleser, Rita, 126–127
  Butts, Susan B., 14–15
  Cummings, Shirlyn, 169
  Davis, L. Shannon, 29–30
  DeMasi, Anne, 20
  Dollar, Beverly, 133
  Dulany, Lissa, 163
  Dunlap, Lou, 41
  Fox, Sharon, 146
  Jackson, Nancy, 48–49
  James, Susan, 121
  Martin, Cheryl, 92
  Rewolinski, Melissa, 157
  Seibert, Rebecca, 99–100
  Slatt, Barbara J., 84–85
  Sullivan, Sally, 35–36
  Thurnauer, Marion, 116
  Torrijos, Grace, 104
  Travis, Tanya, 152
  Vercellotti, Sharon, 55
  Yamashita, Liz, 72–73
  Zappia, Jean, 61
American Chemical Society (ACS)
  DeMasi, Anne and involvement, 18, 20
  Dulany, Lissa, 162–163
  Haynie, Sharon L., 137, 138
  Jackson, Nancy, 43, 47
  PROGRESS program, 9, 181
  Summer Employment for Economically Disadvantaged (SEED), 138

Women Chemists Committee, 176
workforce survey, 1–2
  See also Women Chemists Committee (WCC)
Argonne National Laboratory, Thurnauer, Marion, 111–116
Artman, Diane M.
  advice, 108–109
  balance, 108
  career change, 107
  career path, 105–107
  education, 105
  family, 108
  leadership, 108
  Marketing Director, GrafTech, 106 (photo)
  science as career, 108–109
  success, 107
Association for Women in Science (AWIS), Thurnauer, Marion, 113–114, 115
Attorney, patent, Sullivan, Sally, 31–36
Augsburg College, Keiter, Ellen A., 183
Authors/editors of book
  Davis, Shannon, 181
  Dulany, Lissa, 182
  Erickson, Jacqueline, 182
  Garrison, Arlene A., 182–183
  Hinkle, Amber S., 183
  Keiter, Ellen A., 183
  Kocsis, Jody A., 184
  Leslie, Anne, 184
  Majerle, Rita S., 184
  Piocos, Elizabeth, 185
  Wood-Black, Frankie, 185

B

Balance
    Artman, Diane M., 108
    Barbour, Rachael L., 68
    Bleser, Rita, 126
    Butts, Susan B., 13
    Cummings, Shirlyn, 170–171
    Davis, L. Shannon, 26
    DeMasi, Anne, 19
    Dollar, Beverly, 131–132
    Dunlap, Lou, 40
    Fox, Sharon, 146
    Haynie, Sharon L., 138
    Jackson, Nancy, 48
    James, Susan, 120
    Martin, Cheryl, 90
    Rewolinski, Melissa, 156
    Seibert, Rebecca, 96–97
    Slatt, Barbara J., 83
    Sullivan, Sally, 35
    Torrijos, Grace, 103–104
    Travis, Tanya, 150–151
    Vercellotti, Sharon, 54
    Yamashita, Liz, 71
    Zappia, Jean, 60
    See also Children; Family status
Baldwin-Wallace College, Artman,
    Diane M., 105
Barbour, Rachael L.
    advice, 68
    Battelle Memorial Institute, 65
    career resources and interests, 67–
        68
    early career and education, 63, 65
    family and balance, 68
    preparedness equals success, 66–67
    professional associations, 65–66
    Senior Chemist with Degussa
        Construction Chemicals, 64
        (photo)
Battelle Memorial Institute, Barbour,
    Rachael L., 65
Bayer Corporation, Fox, Sharon, 139–
    146

Bayer Diversity Advisory Council
    (BDAC), Cummings, Shirlyn,
    167
Bayer Material Sciences
    Cummings, Shirlyn, 165–171
    Hinkle, Amber S., 183
    Rand, Lora, 75–78
Biopolymers, Vercellotti, Sharon, 51
Bleser, Rita
    balance, 126
    development, 125–126
    education, 123
    from manufacturing to law, 124–
        125
    reflections and advice, 126–127
    return to manufacturing, 125
    Vice President for Research and
        Development, Mallinckrodt,
        123, 124 (photo)
Brandeis University, Travis, Tanya, 147
Bristol-Myers Squibb Company,
    Yamashita, Liz, 69–73
Burke, Sister Helen, mentor to DeMasi,
    18, 20
Business, Torrijos, Grace, 103
Business climate, Slatt, Barbara J., 82–83
Butts, Susan B.
    balance, 13
    choices and consequences, 14–15
    climate changes, 13–14
    Director of External Technology,
        12 (photo)
    Dow Chemical Company, 11–15
    education, 11–12
    mentors, 14
    success, 14
    transition to management, 12–13

C

California Institute of Technology,
    Sullivan, Sally, 31, 33
Carbohydrate chemistry, Vercellotti,
    Sharon, 51–55

Career development
  Bleser, Rita, 125–126
  Martin, Cheryl, 89–90
Career development tools, Zappia, Jean, 61
Career path
  Artman, Diane M., 105–107
  Barbour, Rachael L., 63, 65
  Bleser, Rita, 124–125
  Butts, Susan B., 12–13
  Davis, L. Shannon, 24
  DeMasi, Anne, 18–19
  Dollar, Beverly, 130–131
  Dulany, Lissa, 159–161
  Dunlap, Lou, 38–39
  Fox, Sharon, 139–141
  Haynie, Sharon L., 135, 137
  James, Susan, 117–119
  Martin, Cheryl, 87, 89
  Rand, Lora, 75–76
  Rewolinski, Melissa, 155
  Seibert, Rebecca, 94–95
  Slatt, Barbara J., 81
  Thurnauer, Marion, 113–114, 115–116
  Torrijos, Grace, 102–103
  Yamashita, Liz, 69, 71
Career resources and interests, Barbour, Rachael L., 67–68
Case Western Reserve University
  Barbour, Rachael L., 66
  Travis, Tanya, 147, 149
Catalyst Award (2002), Cummings, Shirlyn, 165
Central State University, Wood-Black, Frankie, 185
Challenges
  Jackson, Nancy, 46–47
  Thurnauer, Marion, 115–116
Chemical consultant, Dulany, Lissa, 159–163
Chemical professions
  chemists by employer, highest degree, and gender, 7*f*
  demographics, 8, 9*f*

demographics of women and minorities, 1–2
number of chemistry degrees granted in United States, 5*f*
status of women in, 4
unemployed chemists seeking positions and job search restrictions, 6*f*
Chemistry field, Seibert, Rebecca, 97–98
Chemists
  demographics, 8, 9*f*
  employer, highest degree, and gender, 4, 7
  family status, 7–8
Chestnut Hill College, DeMasi, Anne, 18, 20
Children
  Artman, Diane M., 108
  Barbour, Rachael L., 65, 68
  Bleser, Rita, 126
  Butts, Susan B., 13
  Cummings, Shirlyn, 170
  DeMasi, Anne, 19
  Dunlap, Lou, 40
  family status, 7–8
  Jackson, Nancy, 48
  James, Susan, 120
  Rewolinski, Melissa, 156
  Seibert, Rebecca, 96–97
  Sullivan, Sally, 35
  Torrijos, Grace, 103–104
  Travis, Tanya, 150–151
  Vercellotti, Sharon, 54
  Yamashita, Liz, 71
  Zappia, Jean, 57, 60
  See also Balance
Choices
  Butts, Susan B., 14–15
  Seibert, Rebecca, 95, 99–100
Ciba Specialty Chemicals, Zappia, Jean, 57–61
Climate
  Butts, Susan B., 13–14
  Davis, L. Shannon, 25–26
  Dollar, Beverly, 132–133

Dunlap, Lou, 40
Fox, Sharon, 145–146
James, Susan, 119
Martin, Cheryl, 89
Rand, Lora, 78
Slatt, Barbara J., 82–83
Travis, Tanya, 149–150
Yamashita, Liz, 72
Zappia, Jean, 59, 60
College of the Holy Cross, Martin,
    Cheryl, 87
Community focus, Dunlap, Lou, 39–40
Community involvement, Seibert,
    Rebecca, 96
Company president, Vercellotti, Sharon,
    51–55
Competition, Butts, Susan B., 15
ConocoPhillips
    Dollar, Beverly, 129–133
    Wood-Black, Frankie, 185
Consequences, Butts, Susan B., 14–15
Crompton Corporation, Seibert, Rebecca,
    93–100
Cummings, Shirlyn
    balance, 170–171
    Catalyst Award in 2002, 165
    Director of Human Resources,
        Bayer, 165, 166 (photo)
    diversity programs, 167–168
    local initiatives at Bayer's
        Baytown, Texas site, 168–169
    mentoring, 166–167
    personal note, 170
    success, 170–171
    working mothers, 169–170

D

Dana Farber Cancer Institute, Vercellotti,
    Sharon, 54
Davis, L. Shannon
    advice, 29–30
    author, 180 (photo), 181
    career path, 24
    education, 21
    getting started, 21, 23–24
    Leader Process Research and
        Development, 22 (photo)
    leadership, 29
    mentoring, 27–28
    professional development, 28–29
    success, 27
    work climate, 25–26
    work-life balance, 26
Degussa Construction Chemicals,
    Barbour, Rachael L., 63–68
DeMasi, Anne
    advice, 20
    balance, 19
    Burke, Sister Helen, 18, 20
    education and early career, 18–19
    mentors, 18, 20
    Regulatory Specialist, 18 (photo)
    success, 19–20
Demographics
    changes within chemical sciences,
        176–177
    chemists by gender, 8, 9f
    workforce changes, 1–2
Director of Human Resources,
    Cummings, Shirlyn, 165–171
Diversity
    advancement of women, 174–176
    Martin, Cheryl, 91
    programs and Shirlyn Cummings,
        167–168
Dollar, Beverly
    career path, 130–131
    changes in work climate, 132–133
    Chief Intellectual Property
        Counsel, ConocoPhillips, 129,
        130 (photo)
    education, 129
    mentors, 131
    success, 133
    work-life balance, 131–132
The Dow Chemical Company
    Butts, Susan B., 11–15
    Rewolinski, Melissa, 155

Downsizing, Travis, Tanya, 152
Drago, Russell S., Davis, L. Shannon,
  23–24
Dulany, Lissa
  advice, 163
  author, 182
  consultant and writer, 159, 160
    (photo)
  education and career path, 159–161
  networking, 162–163
  opportunity, 161–162
  success, 162
Dunlap, Lou
  advice, 41
  changes in work climate, 40
  community focus, 39–40
  Director of Technology Transfer,
    38 (photo)
  education and early career, 38–39
  success, 41
  work and life balance, 40
Dupont Chemical, Haynie, Sharon L.,
  135–138

E

Eastern Illinois University
  Bleser, Rita, 123
  Keiter, Ellen A., 183
East Technical High School, Travis,
  Tanya, 147
Editors of book
  Hinkle, Amber S., 180 (photo), 183
  Kocsis, Jody A., 180 (photo), 184
  See also Authors/editors of book
Education
  Artman, Diane M., 105
  Barbour, Rachael L., 63, 65
  Butts, Susan B., 11–12
  DeMasi, Anne, 18–19
  Dollar, Beverly, 129
  Dulany, Lissa, 159–161
  Dunlap, Lou, 38–39
  Fox, Sharon, 139–141

Jackson, Nancy, 43, 45
  Martin, Cheryl, 87, 89
  Rand, Lora, 75
  Rewolinski, Melissa, 155
  Seibert, Rebecca, 93
  Slatt, Barbara J., 81
  Sullivan, Sally, 31, 33, 34
  Thurnauer, Marion, 111
  Torrijos, Grace, 102–103
  Vercellotti, Sharon, 53
  Yamashita, Liz, 69
  Zappia, Jean, 57, 59
Emory University
  Dulany, Lissa, 160, 182
  Rand, Lora, 75
Engineering
  percentage of women in fields, 175t
  student enrollment, 3
  workforce, 2–4
England, James, Susan, 117
Environmental responsibility, Haynie,
  Sharon L., 137
Erickson, Jacqueline, author, 180 (photo),
  182
Etzkowitz, Henry, myth of pipeline, 174

F

Family status
  Barbour, Rachael L., 65, 68
  gender differences, 7–8
  See also Balance; Children
Financial planning, Martin, Cheryl, 87,
  89
Foreign assignment, Fox, Sharon, 141–
  142
Fox, Sharon
  advice, 146
  Bayer Corporation, 139, 141
  education and early career, 139–
    141
  foreign assignment, 141–142
  leadership, 145
  mentors, 143–144

professional development, 144
Strategic Planning Consultant,
    Lanxess, 139, 140 (photo)
success, 142–143
work climate, 145–146
work-life balance, 146

G

Gannon University, Seibert, Rebecca, 93
Garrison, Arlene A., author, 180 (photo),
    182–183
Gender
    family status of chemists by, 7–8
    See also Men; Women
George Washington University, Jackson,
    Nancy, 43
Georgia Pacific, Dulany, Lissa, 161
Georgia Southern College, Davis, L.
    Shannon, 21, 23, 181
Georgia State University, Dulany, Lissa,
    160, 182
Germany, Fox, Sharon, 141–142
GlaxoSmithKline Consumer Healthcare
    Erickson, Jacqueline, 182
    James, Susan, 117–121
Golf, Butts, Susan B., 15
Government careers
    Thurnauer, Marion, 111–116
    women, 4, 7
GrafTech International Ltd., Artman,
    Diane M., 105–109
Gray, Harry, mentor for Nancy Jackson,
    46
Greenlee, Winner, and Sullivan law firm,
    Sullivan, Sally, 34–35
Guilt trap, Butts, Susan B., 14

H

Hamline University, Majerle, Rita S., 184
Harvard University, Vercellotti, Sharon,
    54

Haynie, Sharon L.
    American Chemical Society (ACS),
        137
    education, 135
    environmentally responsible, 137
    mentor, 138
    Research Scientist, Dupont
        Chemical, 135, 136 (photo)
    tradition, 135, 137
    work and life, 138
Hinkle, Amber S., editor, 180 (photo),
    183
Human Resources Director, Cummings,
    Shirlyn, 165–171
Hunter College of the City University of
    New York, Sullivan, Sally, 31

I

Industrial careers, women, 4, 7

J

Jackson, Nancy
    balance, 48
    challenges, 46–47
    Deputy Director of International
        Security Center at Sandia
        National Laboratories, 44
        (photo), 45
    development, 47–48
    education, 43
    mentoring, 45–46
    new direction, 45
    research, 45
    success and advice, 48–49
James, Susan
    advice, 121
    balance, 120
    early career, 117–119
    education, 117
    GlaxoSmithKline Consumer
        Healthcare, 117, 118 (photo)

mentors, 120–121
success, 119–120
work climate changes, 119

K

Katz School of Business, Fox, Sharon, 139
Keiter, Ellen A., author, 180 (photo), 183
Kocsis, Jody A., editor, 180 (photo), 184

L

Lanxess, Fox, Sharon, 139, 141
Law
Bleser, Rita, 124–125
Dollar, Beverly, 129–131
Sullivan, Sally, 31–36
Leadership
Artman, Diane M., 108
Cummings, Shirlyn, 169, 170
Davis, L. Shannon, 29
Fox, Sharon, 145
Martin, Cheryl, 91–92
Rand, Lora, 78
Seibert, Rebecca, 97
Slatt, Barbara J., 82
Thurnauer, Marion, 113–114
Leslie, Anne, author, 180 (photo), 184
Louisiana State University, Vercellotti, Sharon, 53
Lubrizol Corporation
Artman, Diane M., 105
Kocsis, Jody A., 184
Travis, Tanya, 147–152

M

McGill University, Leslie, Anne, 184
Majerle, Rita S., author, 180 (photo), 184
Mallinckrodt, Bleser, Rita, 123–127
Management, Butts, Susan B., 12–13

Manhattan College, Zappia, Jean, 57
Manufacturing, Bleser, Rita, 124–125
Marketing Director, Artman, Diane M., 105–109
Martin, Cheryl
advice, 92
career development, 89–90
Director, Financial Planning, 87, 88 (photo)
diversity, 91
education and early career, 87, 89
leadership, 91–92
mentors, 91
success, 90
work climate, 89
work-life balance, 90
Massachusetts Institute of Technology (MIT)
Fox, Sharon, 139
Haynie, Sharon L., 135
Martin, Cheryl, 87
Material safety data sheets (MSDS), DeMasi, Anne, 18
Men
family status, 7–8
number of chemistry degrees in United States by year and gender, 5f
unemployed chemists actively seeking positions, 6f
See also Gender
Mentors
Butts, Susan B., 14
Cummings, Shirlyn, 166–167
Davis, L. Shannon, 27–28
DeMasi, Anne, 18, 20
Dollar, Beverly, 131
Dulany, Lissa, 162–163
Fox, Sharon, 143–144
Haynie, Sharon L., 138
Jackson, Nancy, 45–46
James, Susan, 120–121
Martin, Cheryl, 91
Rand, Lora, 77–78
Rewolinski, Melissa, 156

Seibert, Rebecca, 99
Slatt, Barbara J., 83–84
Thurnauer, Marion, 114–115
tools for career advancement, 173–174
Torrijos, Grace, 104
Travis, Tanya, 151
Vercellotti, Sharon, 55
Zappia, Jean, 61
Minorities
    demographics of chemists by gender, 9f
    science and engineering workforce, 3–4
    workforce demographics of chemical professions, 1–2
Minority Engineering and Industrial Opportunity
Program (MEIOP)
    Travis, Tanya, 149
Monsanto, Davis, L. Shannon, 21–30

N

Nackley, Mary, mentor of Tanya Travis, 151
National Science Board (NSB), report of science and engineering workforce, 2–4
National Science Foundation (NSF)
    DeMasi participating in NSF programs, 18–19
    science and engineering workforce, 3–4
    Vercellotti, Sharon, 53
Networking
    Dulany, Lissa, 162–163
    See also Mentors
Northwestern University, Butts, Susan B., 11
Notre Dame College, Kocsis, Jody A., 184

O

Oak Ridge National Laboratory (ORNL), Dunlap, Lou, 37–41
Ohio Dominican University, Barbour, Rachael L., 63
Ohio State University, Vercellotti, Sharon, 53
Oklahoma State University, Wood-Black, Frankie, 185
Opportunity, Dulany, Lissa, 161–162
Overcoming Challenges Award, Women Chemists Committee (WCC), 181

P

Patent attorney, Sullivan, Sally, 31–36
Pfizer Global Research and Development, Rewolinski, Melissa, 153
Pharmacia and Upjohn, Rewolinski, Melissa, 155
Piocos, Elizabeth, author, 185
Pipeline myth, Etzkowitz, Henry, 174
Polytechnic Institute of New York, Zappia, Jean, 57
President, Vercellotti, Sharon, 51–55
Proctor & Gamble Company
    Piocos, Elizabeth, 185
    Slatt, Barbara J., 79–85
Professional associations
    advancement of women, 176
    Barbour, Rachael L., 65–66
    Dulany, Lissa, 162–163
    Martin, Cheryl, 90
    Slatt, Barbara J., 81
    Thurnauer, Marion, 113–114
Professional development
    Bleser, Rita, 125–126
    Davis, L. Shannon, 28–29
    Fox, Sharon, 144
    Jackson, Nancy, 47–48
    Martin, Cheryl, 89–90

Rewolinski, Melissa, 156
Slatt, Barbara J., 84
Thurnauer, Marion, 115
Yamashita, Liz, 71–72
Professional experience, Vercellotti,
  Sharon, 53–54
PROGRESS program, American
  Chemical Society (ACS), 9, 181

R

Rand, Lora
  career path, 75–76
  education, 75
  leadership, 78
  mentoring, 77–78
  success, 76–77
  VP plastics manufacturing, 76
    (photo)
  work climate, 78
Reflection
  Bleser, Rita, 126–127
  Sullivan, Sally, 35–36
Regis University, Wood-Black, Frankie,
  185
Regulatory field
  DeMasi, Anne, 18
  James, Susan, 118–119
  Yamashita, Liz, 71–72
Reorganization, Travis, Tanya, 152
Research, Jackson, Nancy, 45
Rewolinski, Melissa
  advice, 157
  balance, 156
  Director Chemical Research and
    Development, 153, 154 (photo)
  education and early career, 155
  mentoring, 156
  professional development, 156
  teamwork, 153
  work and life, 156
Rice University, Rewolinski, Melissa,
  155
Rohm and Haas Company

DeMasi, Anne, 17–20
Martin, Cheryl, 87–92
Role models
  career advancement, 173–174
  See also Mentors
Rutgers University–Newark, Erickson,
  Jacqueline, 182

S

St. Louis University, Bleser, Rita, 124–
  125
Sandia National Laboratories, Jackson,
  Nancy, 45–48
Science
  percentage of women in fields, 175t
  student enrollment, 3
  workforce, 2–4
Science Careers in Search of Women,
  Thurnauer, Marion, 113
SEED. *See* Summer Employment for
  Economically Disadvantaged (SEED)
Seibert, Rebecca
  advice, 99–100
  balancing career and home life, 96
  career path, 94–95
  chemistry yesterday, today, and
    tomorrow, 97–98
  choices, 99–100
  education, 93
  mentoring, 99
  mother and professional, 96–97
  sacrifice or opportunity, 95
  success, 98
  Technology Manager for Crompton
    Corporation, 94 (photo)
  traits of true leader, 97
Shell Development, Rewolinski, Melissa,
  155
Slatt, Barbara J.
  advice, 84–85
  balance, 83
  business climate, 82–83
  education and early career, 81

Manager Corporate
R&D/Corporate External
Relations, 80 (photo)
mentoring, 83–84
Proctor & Gamble Company, 79,
81
professional development, 84
success, 81–82
Solar Environmental Services, Inc.,
Torrijos, Grace, 101–104
Solutia/Monsanto, Davis, L. Shannon,
21–30
Strategic planning consultant, Fox,
Sharon, 139, 141
Success
Artman, Diane M., 107
Barbour, Rachael L., 66–67
Butts, Susan B., 14
Cummings, Shirlyn, 170–171
Davis, L. Shannon, 27
DeMasi, Anne, 19–20
Dollar, Beverly, 133
Dulany, Lissa, 162
Dunlap, Lou, 41
Fox, Sharon, 142–143
Jackson, Nancy, 48–49
James, Susan, 119–120
Martin, Cheryl, 90
Rand, Lora, 76–77
Seibert, Rebecca, 98
Slatt, Barbara J., 81–82
Thurnauer, Marion, 116
Torrijos, Grace, 104
Travis, Tanya, 151
Vercellotti, Sharon, 54
Yamashita, Liz, 72
Zappia, Jean, 60
Sullivan, Sally
adjusting to change, 34
advice and reflection, 35–36
balance, 35
branching out, 33–34
education, 31, 33, 34
patent attorney, 32 (photo)
research, 33

shaping work environment, 34–35
Summer Employment for Economically
Disadvantaged (SEED)
American Chemical Society (ACS),
138
Leslie, Anne, 184
Synergen, Sullivan, Sally, 33

T

Teamwork, Rewolinski, Melissa, 153
Thurnauer, Marion
advice, 116
Argonne National Laboratory, 112
(photo)
challenges, 115–116
development, 115
education, 111
mentoring, 114–115
path to leadership, 113–114
success, 116
Women in Science Program,
113
Torrijos, Grace
advice, 104
business, 103
education and early career, 102–
103
mentors, 104
owner and president, Solar
Environmental Services, Inc.,
101, 102 (photo)
success, 104
work-life balance, 103–104
Travel awards program, Women
Chemists Committee (WCC), 181
Travis, Tanya
advice, 152
balancing work and family, 150
difficult times, 152
education, 147, 149
Lubrizol Corporation, General
Manager, 148 (photo), 149
mentoring, 151

Minority Engineering and
Industrial Opportunity Program
(MEIOP), 149
mother and professional, 150–151
success, 151
supportive employer, 149–150
work climate, 149–150

U

UCB (Belgian chemical/pharmaceutical
company), Dulany, Lissa, 161, 182
United Kingdom, James, Susan, 117
University of Alaska, Anchorage,
Torrijos, Grace, 102–103
University of Arkansas, Vercellotti,
Sharon, 53
University of Central Oklahoma, Wood-
Black, Frankie, 185
University of Chicago, Thurnauer,
Marion, 111
University of Cincinnati, Artman, Diane
M., 105
University of Delaware, Erickson,
Jacqueline, 182
University of Denver College of Law,
Sullivan, Sally, 34
University of Florida, Davis, L. Shannon,
21, 23–24, 181
University of Illinois, Slatt, Barbara J., 81
University of Illinois at Urbana-
Champaign, Keiter, Ellen A., 183
University of Michigan, Butts, Susan B.,
11
University of Minnesota, Majerle, Rita S.,
184
University of New Haven, Seibert,
Rebecca, 93
University of Oklahoma, Dollar, Beverly,
129
University of Pennsylvania, Haynie,
Sharon L., 135
University of Pittsburgh, Seibert,
Rebecca, 93

University of Rochester, Yamashita, Liz,
69
University of Tennessee
Dunlap, Lou, 38, 41
Garrison, Arlene A., 182–183
University of Texas
Fox, Sharon, 139
Jackson, Nancy, 43, 45f
Rand, Lora, 75
University of Tulsa, Dollar, Beverly, 129
University of Utah
Hinkle, Amber S., 183
Leslie, Anne, 184
University of Virginia, Dulany, Lissa,
159–160, 182
University of Washington, Hinkle,
Amber S., 183

V

Valian, Virginia, advancement of
women, 175–176
Vercellotti, Sharon
advice, 55
education and professional
experience, 53–54
founder, owner and president of V-
LABS, INC., 51, 52 (photo)
mentoring, 55
success, 54
work and life balance, 54
Villanova University, DeMasi, Anne, 19
V-LABS, INC., Vercellotti, Sharon, 51–
55

W

Washington University Law School,
Bleser, Rita, 125
Women
Bayer and working, 169–170
chemists by employer, highest
degree, and gender, 7f

demographics, 8, 9f
employment categories, 4, 7
family status, 7–8
number of chemistry degrees in United States by year and gender, 5f
percentage in science and engineering fields, 175t
science and engineering workforce, 3–4
serving as role models, 9
status in chemical professions, 4
unemployed chemists actively seeking positions, 6f
workforce demographics of chemical professions, 1–2
See also Gender
Women Chemists Committee (WCC)
advancement of women, 176
authors/editors of book, 180 (photo), 181–185
finding successful women for role models, 9
interviewing successful women chemists, 179
issues, 2
supporting women in chemical sciences, 179–181
Thurnauer, Marion in newsletter, 115
travel awards program, 181
WCC Overcoming Challenges Award, 181
Women in Science Program, Thurnauer, Marion, 113
Wood-Black, Frankie, author, 185
Work climate. See Climate
Work environment, Sullivan, Sally, 34–35
Workforce, science and engineering, 2–4
Workforce demographics, women and minorities in chemical professions, 1–2

Workforce diversity, advancement of women, 174–176
Working mothers,
Cummings, Shirlyn, 169–170
See also Balance; Children
Work-life balance. See Balance
Writer
Dulany, Lissa, 159–163
See also Authors/editors of book

X

Xavier University, Slatt, Barbara J., 81

Y

Yamashita, Liz
advice, 72–73
balance, 71
climate changes, 72
education and career, 69, 71
Group Director, Global Regulatory Sciences, 70 (photo)
regulatory field, 71–72
success, 72

Z

Zappia, Jean
advice, 61
career development tools, 61
Ciba Specialty Chemicals, 57
education, 59
mentoring, 61
success, 60
VP Plastics Additives, 58 (photo)
work climate, 59
Zozulin, Alex, Davis, L. Shannon, 23